Sustainable Cities for the Third Millennium:
The Odyssey of Urban Excellence

Voula P. Mega

Sustainable Cities for the Third Millennium: The Odyssey of Urban Excellence

 Springer

Voula P. Mega, after a special authorisation by the European Commission
PhD in City & Regional Policy and Planning
Brussels
Belgium
voula.mega@ec.europa.eu
Voula.Mega@post.harvard.edu

ISBN 978-1-4419-6036-8   e-ISBN 978-1-4419-6037-5
DOI 10.1007/978-1-4419-6037-5
Springer New York Dordrecht Heidelberg London

Library of Congress Control Number: 2010934906

Printed on acid-free paper

Springer is part of Springer Science+Business Media (www.springer.com)

In the loving memory of
Panagiotis & Alexandra Megas
who dreamt great dreams
and first taught me
the greatness, integrity and beauty of our world

Disclaimer: "The views expressed are purely those of the writer and may not in any circumstances be regarded as stating an official position of the European Commission. The European Commission cannot be held liable for any use of the information contained in this document."

"The city is built politics" (Aristotle, IV ct. BC)

"Crossroad of people, the city is also a crossroad of influences" (A. Rimbaud "Illuminations", 1875)

"The global effort for sustainability will be won, or lost, in the world's cities, where urban design may influence over 70 percent of people's Ecological Footprint"(Global footprint network, 2008)

"When you set out on your journey to Ithaca, pray that the journey is long, full of adventure, full of knowledge"(C. Cavafy Ithaca, 1911)

Preface

At the dawn of the third millennium, planet Earth entered a zone of turbulence. The 2008 crisis added economic uncertainty to the threat of global warming and extreme events such as droughts, floods and cyclones, the persisting crisis of poverty and the spectrum of pandemics and terrorism. Against this global landscape in an era of fragility, cities, already sheltering more than half of humankind, appear as Janus-faced realities, the best and worst of places, vulnerable but still full of hope and will to overcome the crisis of societal values and progress in the path of sustainable development.

This book addresses the most critical challenges for cities, humanity's collective masterpieces in danger, and analyses breakthrough responses for sustainable development, a globalisation with human face and the transition to inclusive post carbon communities. The ultimate wish is that experts, city planners, decision-makers and citizens in search of sustainable cities could find here some sources of information and inspiration to enhance the immense possibilities of cities and embrace the best possible trajectories of change.

Cities have crossed some sensitive boundaries and reached critical biophysical thresholds. Many of them are on the brink of potentially irreversible change. Both perpetrators and victims of unbridled consumption, they have a preeminent role to play for recreating a dynamic balance between the quantity and quality of development and a harmonious symbiosis among nature, humans and artefacts. Sustainable development should be at the heart of the recovery if the death knell of prosperity is to be prevented. Furthermore, sustainable development could give the exiting from the crisis a sense of purpose.

This book advocates for strong sustainability, implying that not only the overall urban capital should not be diminished, but also each of its precious components. The environmental (including the life support systems), natural (the physical assets transacted in markets), human (health, education and well-being), social (networks, traditions, culture), including political (the ability of citizens and communities to take control of their future), and human-made (financial and built environment) capital have to be preserved and transmitted.

The battle for sustainable development will be won or lost in (or because of) cities. The first reason is simply their sheer size. The extraordinary concentration of people and activities are the very essence of cities. Furthermore, cities are the only places where people and resources congregate at a point beyond which synergetic effects become more important than the simply additive ones.

Although general world figures conceal large differences across countries and regions and should be considered with caution, cities host more than 50% of the inhabitants of the planet, consume more than two thirds of its resources and produce more than two thirds of the global greenhouse emissions. However, the consumption of resources and production of emissions per capita is lower for urban dweller than for the rest of the inhabitants of the planet. Cities can drive efforts towards sustainability in a more efficient manner than the rest of the globe.

The cultural and social reasons for sustainability to be won or lost in cities are equally important. Cities, beehives of creativity and innovation, have the capacity to engender new values and concepts and introduce structures and patterns, which are quickly disseminated to the rest of the world. There are also fundamental political reasons. Cities are organised societies that always promoted local democracy and this is a major precondition for advancing towards sustainable development. Finally, cities bear also witness of the will to change and progress. An extraordinary number of symbolic events, awards and emblematic projects reveal a mine of ideas, energy invested in action and a potential still to be realised.

This book highlights that to win the battle for sustainable development, cities have to go through a revolution in concepts and actions and sheds light to pathways of excellence and change embraced by pioneer cities. It addresses all dimensions of urban capital, which have to be preserved and enhanced for sustainability, synthesised in chapters that correspond to policy fields and themes. Each part invites for trans-border interdisciplinary approaches and tries to answer two interlinked questions: "Where the forefront of excellence lies" and "What cities do to approach and overcome it".

The first chapter examines the world and urban demographic and socio-economic dynamics. The European archipelago of cities bears witness of more than two decades of visions and actions for urban sustainability. European cities can offer many lessons on their balance-seeking processes in search of sustainable development, extending into all areas of local decision-making.

In a context of global crisis, progress towards sustainability requires major breakthroughs and paradigms shifts. More and higher value has to be derived from knowledge, an inexhaustible resource with a high empowerment potential. Cities have to invest more creatively in their intangible assets and breed innovations needed by their body, with its blood, its vital organs and nerves, but also their mind, with its concepts and notions, and their soul, with its emotions and capacity to wonder. Sustainability means, first and foremost, sustaining ability to innovate and progress.

The second chapter focuses on key elements of the urban ecosystems and principles and actions inaugurating new models and patterns of consumption and production. Cities are complex and dynamic, open and vulnerable ecosystems, the only human and socio cultural ones. By analogy to the biological concentration of species in a mutually supportive environment, humans come together in places that optimise their reciprocal benefits. As socio-bio-spaces, cities require an extraordinary array of material and labour, they have most complex metabolisms and produce an equally remarkable array of products, waste and emissions. Their role in the transition to a low-, zero- and post-carbon society and economy is crucial.

Climate change, the end result of billions of everyday resolutions taken in local environments, will be decided in cities or because of the values and models that cities disseminate in the world. Urban policies should consider all measures in order to radically reduce the ecological and carbon footprints and the environmental shadow of cities. Resource (including waste) management, quality of land, water and air as well as noise, are of extreme importance to cities that become laboratories of ecological innovation and provide inspiring examples.

Eco-cities are not simply cities with parks, grass roofs, eco-performing public transport, timber frame constructions, improved energy systems and water cycles. They are cities that pertinently integrate environmental, social and economic policy objectives in all their policies and search the most advanced technologies and practices to improve their overall performance, based on strong citizen participation. From eco-buildings to eco-districts and from eco-cities to the only one planet, actions have to be proportionate to the challenges. Throughout the book, the prefix "eco-" is used for both the ecology and the economy and is suggested as symbol for double dividend projects, good for both the environment and the economy.

The transport sector, especially road transport, is source of the highest increasing emissions. Transport has vital links with all dimensions of sustainable development, but it experiences problems to be on a sustainability path. The third chapter offers an insight into urban mobility and accessibility patterns and models and the challenges for profound change. The car is still by far the dominant transport mode in cities even if its supremacy is challenged and citizens ask for better transport services. Urban arteries often get blocked by traffic, pollution and emissions and may lead to urban thrombosis in cities, if a better balance between individual and collective transport is not urgently stricken.

However, cleaner transport options and services exist and can be attractive. Highly eco-performing public transport can be the backbone of sustainable public transport services. Walking and cycling are the only truly sustainable transport means. Car-free districts and cities should be rooted in the local culture and respond the citizens needs and concerns. Mobility weeks and symbolic events can boost the demand for competitive and citizen- and environment-friendly urban transport.

Cities and governments have to ensure secure, competitive, clean and affordable energy supply. They also have a responsibility to influence demand patterns linked to consumer behaviour and society lifestyles. Forward-looking studies highlight that, in the absence of bolder policies, the energy future will be unsustainable in all its dimensions, environmental, economic and social. But this can and should be altered. The "20%/20%/20%" European targets for 2020 can be mobilising for cities, which have often more ambitious plans to improve energy efficiency, reduce greenhouse emissions and increase the share of renewable energies in their energy portfolio.

The fourth chapter examines renewable energy sources and cleaner energy options and technologies and highlights worthy efforts in cities to realise the potential of solar and wind energy, bioenergy, fuel cells and hydrogen. It presents the ethical questions linked to the access and use of energy and reviews citizen perceptions, explored by recent European energy polls. Finally, it argues that nuclear fusion may be a competitor for renewable energies in some decades from now.

The future of cities depends largely on their capacity to generate and distribute wealth and the quality of life that offer to citizens. Cities are both the creators and the mediators of the economic dynamics that are released by the mutual reinforcement of activities clustering together. They are not only the theatres of socio-economic operations, but powerful generators of the wealth of the nations. In times of crisis, the search of sustainable growth, in balance with the other components of sustainability, is critical. Multi-win innovations are essential for creating new assets, especially out of liabilities.

The fifth chapter examines the competitiveness of cities, propellers of growth and the assets that have to be enhanced for sustainability. It argues that smart and green entrepreneurship and employment creation is crucial for going up the urban value chain. Innovative partnerships with eco-businesses are cardinal for reconciling green growth benefits with long term sustainability goals. Quality of life and access to resources and knowledge are key factors of competitiveness, as highlighted by international surveys, assessments and rankings. The chapter ends at the competitive edge of the cities, their lungs and brains, the green and grey parks.

The sixth chapter sheds light on the social capital of cities, as theatres of traditions, wills and wishes. Cities are schools for respecting and embracing difference and learning life in society. Social justice is a prime criterion for judging urban sustainability. Unequal distribution of resources has draining effects on the vitality of urban economies. It may be source of both unsustainable lifestyles and obstacles to cultural change. Equity and social justice, public health and well-being, eco-responsible housing and living conditions, safety and solidarity touch the very heart of cities and impact their potential for sustainability.

Deprived suburban areas, the back scene of most urban theatres, need special attention, as they are the places where most disadvantaged and excluded citizens congregate. Citizens can play a major role in creating vital urban spaces and fulfilling neighbourhoods out of degraded spaces. Participation of youth and women participation in projects can especially extend the limits of the possible. Education is the key and most productive long-term investment. Best practices and awards provide an invaluable opportunity to promote projects most worthy of recognition and praise and share lessons and messages.

The seventh chapter examines the role of sciences, culture and the arts for urban sustainability. Cities are incomparable heritages and beehives of creativity, shared sights, sounds and scents. Knowledge cities and local helices invest in innovation and provide valuable models for targeting investment in science and technology in order to become word centres of excellence. Heritage and culture define urban identity. Citizens project their hopes and desires into the urban reality and legend, while arts in the city represent the ultimate expression of collective intelligence.

The eighth chapter highlights models and actions for urban renaissance, vital to support the process of permanent renewal that sustainable development implies.

Land use planning and transport are fundamental and interrelated instruments for the sustainable regeneration of cities, of their physical parts and of their extraordinary diversity. Many cities demonstrate that the intensification and consolidation of the urban fabric can prevent uncontrolled urban sprawl. Harbour cities face special challenges and many urban waterfronts undergo radical transformational change.

Dignified public spaces can promote collective life and local democracy and bring added value to places. Symbolic and structural projects can become emblems of the urban future. The chapter underlines that strategic planning for urban renaissance has to address not only the three spatial dimensions but also the time dimension of cities. Time is also a scarce and most precious resource. Local time plans can enhance the capacities of cities like chronotopes and improve quality of life.

Citizens are the political stakeholders of cities. They have the right to information and the duty to exercise the democratic monitoring of local policies. Since the time of the Athenian city-State, the permanence of cities in time has been related to their capacity to promote open democracies. Active citizenship means participation in and responsibility for decisions on the future of a city.

The ninth chapter examines the emergence of new models of citizen participation in accountable cities, a sinequa non condition for sustainability. Innovative partnerships can maximise the potential of synergies, enrich content and methods of cooperation and serve as catalysts of transformational change. Institutional alliances are enriched with a variety of participatory schemes. World coalitions of cities and citizens, both from the developing and the developed world, can play a major role in addressing global common challenges and achieving the millennium goals for the renaissance of the planet.

Sustainability indicators can take the pulse of a city and provide insight regarding progress towards sustainable development. They may serve as a yardstick and be instrumental for policy making and for assessing policy implementation. But, as Albert Einstein formulated, "not everything that can be counted counts and not everything that counts can be counted." Qualitative indicators are often necessary to complement quantitative frameworks.

The last chapter insists on the importance of a sustainability policy framework for the development of indicators and presents some influential practices with indicators and targets. It finally proposes a set of headline policy-significant environmental, economic and socio cultural indicators for cities, in line with the principles of the European Charter of Sustainable Cities.

The book offers a 360° view of the challenges facing cities in terms of growth, governance, sustainability, social inclusion, science, culture and the arts, urban renaissance and planning. It presents responses mainly from European cities, but tries to approach the moving forefront of urban excellence in the world and distil useful messages for all cities wishing to adopt bolder and better policy initiatives.

Each chapter focuses on an essential dimension of the sustainable city and presents, in a nutshell, visions and actions of cities, from the European and international horizon that herald meaningful far-reaching change. A special effort has been made for the presented actions to be always win-win, and if possible multi-win and contribute to, at least two, dimensions of urban sustainability.

Homer's *Odyssey* has been the greatest poem of all time inviting the opening of the world. The destination is also the journey for cities striving to raise the bar of excellence and sail towards sustainable development in all its vital seas and critical oceans. Excellence is a dynamic notion based on continuing weathering of storms and perpetual surpassing. This Odyssey invites a noble emulation to harness innovation as a springboard for progress.

The writing of this book was finished just before the Odyssey of cities reached Copenhagen in December 2009. Cities promise to be very active in cooperating for better negotiations on climate change. This book tried to capture something of their hopes and wishes to transform the current crisis of values and models into an opportunity and lead concerted action to sustainable futures. May cities grasp this moment of possibility and results exceed their wishes.

Brussels, Belgium   Voula P. Mega

Foreword by Professor Sir Peter Hall

There are many books flooding from the presses on the theme of sustainable urbanism. There is none, from my knowledge, that is quite like Voula Mega's Sustainable Cities for the Third Millennium. None manages to include anything near the amazing variety of case studies that she summarises and quotes in these pages: examples of places that in different ways, over recent years, have managed to achieved a special degree of urban excellence. And, since the devil is always in the detail, this makes her book a special kind of compendium, of unique value to urban researchers and policymakers.

Part of this richness, it must be clear, comes from the author's personal familiarity with the places she describes, stemming from her career as an urban expert within the EU bureaucracy – a career that required a special authorisation and a personal disclaimer from her: these are her judgments, not those of the European Commission. That deep personal knowledge is clear, in some cases, from the wonderful watercolours that introduce each chapter. And significantly, the majority of those illustrations, as of the cases, refer to European cities.

Is her book Eurocentric, then? Yes, in one sense but no in another. The bias, if you can call it that, arises quite simply for two reasons. The cities she instances, in very many cases recurrently, just happen to be the best-practice cities in the world, visited and revisited by professional urban pilgrims from every continent. Stockholm and Copenhagen and Helsinki, Amsterdam and Berlin and Freiburg, London and Dublin, Lyon and Orvieto and Segovia, and many more, set the standards which other cities seek to follow. And, as she shows, EU initiatives have powerfully assisted this process, by encouraging cities both to compete and to cooperate in a constant effort to raise standards.

But her focus is much more than European: it is global. UN as well as EU initiatives play an important role in these pages. And cities across the world, from Tokyo to Curitiba, have also been successfully struggling to raise the quality of life for their people, on multiple dimensions: economic, environmental, cultural. At the start of the third millennium, to a degree never before known, we all inhabit a single global world and take part in a global discourse – to which Voula Mega's new book makes such a significant contribution.

Professor Sir Peter Hall
Bartlett Professor of Planning and Regeneration

Foreword by Professor Athena Linos

It is a great pleasure for me to contribute a foreword to Voula Mega's extraordinary new book on sustainable cities at the beginning of a millennium full of fears and hopes. This book goes further in the search for excellence demonstrated in "Sustainable development, energy and the city".

The merits of the book are multiple. I am sure that interested readers, from the most specialised experts and professionals to city administrators and civil society, will find here the most advanced concepts, challenges and responses that science, innovation and citizen action can provide. Once more, Voula Mega makes science accessible to all for the benefit of science and society alike. She strikes the right balance between "broad horizon approach" and "depth of analysis", general considerations and particular cases and offers an exceptional array of very best practices from cities worldwide.

The book examines urban sustainability in all its dimensions, environmental, socio economic, political and cultural, and asks "What is at stake", "What are the factors of success", "How best can cities address it", "What can inspire cities", "What is the spark of innovation and excellence", "How can cities and citizens drive change and become actors in their future?" It perfectly responds to the call for "Better city, better life", the title of the 2010 World Exhibition in Shanghai, first world Expo in a developing city yet global power. In the aftermath of the Copenhagen Summit, the expectations for overcoming the crisis of economic and environmental values concentrate in cities, propellers of innovations and green growth.

All those who followed the lectures given by Voula Mega in various world Universities, will enjoy learning more about the forefront of innovation and excellence in cities. Her latest book increases our shared awareness of cities as "collective masterpieces in danger" and invites to raise the bar of excellence towards "Better Cities for a better World".

Athens University   Professor Athena Linos

Acknowledgements

If writing a book is a journey, writing this book has been a particularly enriching one. Many dear colleagues and friends advised on the navigation through the islands of the cities of the future. I thank them from heart for all the valuable compasses that have provided.

In trying to increase awareness on cities as "collective masterpieces in danger" and their immense possibilities to turn adversity into opportunity, the book expanded like the universe and the galaxies, to take on board a myriad of worthy projects. Thank you to all those who suggested some precious quasars, shining distant stars, to shed further light to the issues.

Among those who repeatedly honoured me with their trust and insight, without having never being publicly thanked, a particular gratitude is due to Professor Sir Peter Hall and Professor Athena Linos, two bright examples throughout my career.

Great thanks to Anna-Carin Krokstäde-Kneip and Michel Kneip who, once more, reviewed attentively the final manuscript and formulated most precious remarks. The last revision of the manuscript was done in Bangkok, an extraordinary urban laboratory, in autumn 2009 and benefited much from the most inspiring discussions with Dr. Suthawan Sathirathai.

Last but not least, all my thanks to Melinda Paul, Senior Editor with Springer, and her team for their warm professionalism and kind efficiency.

Acronyms

ACE	Architects' Council of Europe
ACRR	Association of Cities and Regions for Recycling
ARAU	Atelier de Recherche et d'Action Urbaine, Brussels
BBP	Better Buildings Partnership, Toronto
CCI	Clinton Foundation's Climate Initiative
CCP	Cities for Climate Protection
CEMR	Council of European Municipalities and Regions
CERES	Coalition of Environmentally Responsible Economies and Societies
CHP	Combined Heat and Power
COMEST	UNESCO's World Commission on the Ethics of Scientific Knowledge and Technology
COP	Conference of the Parties (COP 15 in Copenhagen, December 2009)
CSD	Commission on Sustainable Development
CSR	Corporate Social Responsibility
ECCP	European Climate Change Programme
ECMT	European Conference of Ministers of Transport
ECSC	European Coal and Steel Community
EEA	European Environment Agency
EFTA	European Free Trade Association (Iceland, Liechtenstein, Norway)
EFUS	European Forum for Urban Safety
EMAS	Eco-Management and Audit Scheme (The integration of energy efficiency in EMAS led to the E2MAS)
EPBD	Energy Performance of Buildings Directive
ESCT	European Sustainable Cities and Towns
ETS	Emission Trading Scheme
EU	European Union
EU15	European Union of 15 member States
EU25	European Union of 25 member States (1.5.2004–1.1.2007)
EU27	European Union of 27 member States (since 1.1.2007)

EURATOM	European Atomic Energy Community
EUREC	European Renewable Energy Centres
EWEA	European Wind Energy Association
FMCU-UTO	World Federation of United Cities
G7	The finance ministers and Central Bank governors of Canada, France, Germany, Italy, Japan, United Kingdom and United States
G8	The above plus Russia
G20	The finance ministers and Central Bank governors of Argentina, Australia, Brazil, Canada, China, France, Germany, India, Indonesia, Italy, Japan, Mexico, Russia, Saudi Arabia, South Africa, South Korea, Turkey, United Kingdom, United States and the European Union, represented by the Council and the European Central Bank
GDP	Gross Domestic Product
GEF	Global Environmental Facility
GHG	Greenhouse Gases (carbon dioxide (CO_2), methane (CH_4), nitrous oxide (N_2O) and the three main fluorinated gases)
GPP	Green Public Procurement
HDI	Human Development Index
HIV/AIDS	Human Immunodeficiency Virus/Acquired Immune Deficiency Syndrome
IAEA	International Atomic Energy Agency
ICLEI	International Council for Local Environmental Initiatives
IEA	International Energy Agency
IFHP	International Federation of Housing and Planning
IP	Intellectual Property
IPCC	Intergovernmental Panel on Climate Change
ISOCARP	International Association of City and Regional Planners
ITER	International Thermonuclear Experimental Reactor
ITF	International Transport Forum
LZC	Low/Zero Carbon
NGO	Non-governmental Organisation
OECD	Organisation for Economic Co-operation and Development
OPEC	Organization of the Petroleum Exporting Countries
PPP	Purchase Power Parity
RES	Renewable Energy Systems
RES-E	Electricity from Renewable Energy Sources
SME	Small and Medium-sized Enterprise
UCLG	United Cities and Local Governments
UITP	Union Internationale des Transports Publics

UNCED	UN Conference on Environment and Development (Rio de Janeiro, 1992)
UNDP	UN Development Programme
UNECE	UN Economic Commission for Europe
UNEP	UN Environmental Programme
UNESCO	UN Educational, Scientific and Cultural Organisation
UNFCCC	UN Framework Convention on Climate Change
UNPF	UN Population Fund
WBCSD	World Business Council for Sustainable Development
WEF	World Economic Forum
WHO	World Health Organisation
WMCCC	World Mayors' Council on Climate Change
WMO	World Meteorological Organisation
WSSD	World Summit on Sustainable Development
WTO	World Trade Organisation

Contents

List of Watercolours

Cover design: Sustainable cities in the lap of nature

The cover design is from the author's triptychs on "Cities in the lap of the nature". The watercolours that unite chapters are from the author's collection "Solar cities"

Watercolour 1
The Archipelago of Cities

The archipelago of cities

Chapter 1
Cities in an Era of Fragility

Since the early years of the third millennium, the urban dwellers of planet Earth have outnumbered the rural ones. Forward-looking studies have highlighted the vital role of cities in the future world. This scene-setting chapter examines the world and urban dynamics and focuses on the European archipelago. It reviews, in a nutshell, 20 years of visions and actions for urban sustainability, a dynamic creative balance-seeking process, extending into all areas of local decision-making.

In a context of a global crisis, progress towards sustainability requires major breakthroughs and paradigm shifts. For significant change, strong sustainability, implying the preservation of each component of the urban capital, is needed more urgently than ever. The environmental (including the life support systems), natural (the physical assets transacted in markets), human (health, education and well-being), social (networks, traditions and culture), including political (the ability of citizens and communities to take control of their future), and human-made (financial, built environment) capital have to be preserved and transmitted. More value has to be derived from knowledge, an inexhaustible resource with a high empowerment potential. Sustainability means, first and foremost, sustaining ability to innovate and progress.

1.1 Cities in an Urbanising Planet

1.1.1 At the Crossroads of Cities

At the beginning of the century, humanity is standing at the crossroads of cities. For the first time in history, the urban dwellers of the planet outnumber the rural ones. The global urban population exceeded the 50% mark in 2007 (The Economist 2007). According to the last UN population estimate, 60% of the world population is expected to live in urban areas in 2030 and almost all of the growth is expected to occur in the urban less developed world. Beyond the demographic growth, urbanisation is an ultimate cultural process and a key issue for sustainable development (UN 2007; UN/HABITAT 2006).

V.P. Mega, *Sustainable Cities for the Third Millennium: The Odyssey of Urban Excellence*, DOI 10.1007/978-1-4419-6037-5_1,

In the first years of the millennium, the global economy achieved a growth record. With an average growth rate of 3.2% during the years 2000–2005, it grew more than in any 5-year period since the Second World War and in spite of a number of geopolitical shocks, including the terrorist attacks of 11 September 2001, the wars in Afghanistan and Iraq and the breakdown of the WTO Doha round (OECD 2007b). After the financial crisis caused by excessive borrowing hit the planet in 2008, it became obvious that this growth model founded on narrow-profit considerations and short-term values was very vulnerable and uncertain and fragile.

Many voices during the previous years had raised alerts about the absurdity of this status quo and the fragility of this model of values and asked for the rescue of a civilisation in trouble, especially in relation to climate change. Brown (2006) asked for the urgent replacement of the fossil-fuel-based, throwaway economy with a new economy powered by abundant sources of wind and solar energy.

"If within every crisis lies an opportunity", the crisis that started in 2008 may bring about a great chance for reconsidering values and making progress towards sustainable development. Authoritative voices have asked for a new Green Deal to give the exiting from the crisis a sense of purpose. Forging a consensus on halting greenhouse gas emissions could herald an era of progress towards a decarbonised prosperous and inclusive society, a compelling green deal (Worldwatch Institute 2009a; Renner et al. 2009). The creation of the G20 in the increasingly multipolar world and the coordination of efforts to overcome the crisis should be accompanied by bold decisions for a transition to sustainable development.

Urban societies are champions of excess borrowing. Cities occupy a very small percentage of the surface of the planet but consume an extraordinarily large share of its resources (Rogers and Gumudjian 1994). They have, like Janus, two faces; defined as "the best and the worst of places" (Charles Dickens), they become more ambivalent; they include but also exclude, assemble but also divide.

Cities are human nests and havens, but equally human jungles and places of social confrontation and conflict. Their environmental and carbon footprints have increased beyond rational limits. The search for sustainable urban development is more intense than ever, in Europe and in the rest of world. Many cities wish to be leaders in the transition to sustainable development and propose new visions and structures for recreating humanity's harmony with the environment.

Cities evolve as a result of their assets, the policies to enhance them and the means used to this end. The passage from the walled feudal city to metropolises and to metapolises embodies the technological progress from agrarian to industrial production and the information society. From the eotechnical phase, marked by a shift in the energy source from humans to animals and machines, to the paleotechnical phase, based on coal and iron, and from the neotechnical phase of assembly line production to twenty-first century technologies and innovations, cities kept constantly transforming and often surprising themselves.

Cities embody the values of the civilisations which created and shape them, diffuse patterns and provide the springboard to new eras. According to Plato, the aim of a good city is to lead citizens towards a happy life. Each city is a *bonum commune*, a living organism and a technical construction. Levi-Strauss called it the "human invention par excellence." Mumford spoke about the city as "the form and symbol of an integrated social relationship", and stated that "the test of a great city is the life it makes possible for its citizens." Jacobs (1969) defined cities as places that continuously generate, in an ongoing way, their economic growth from their endogenous resources and from the "disorderly order" of human interactions (ENA Recherche 1996).

Cities are theatres of civilisation, schools of abilities and values, beacons of culture and temples of learning about life in society, citizen duties and rights. They have been defined as places in which the human genius is expressed, a palette of collective potential yet to be realised. In the fourth century BC, Aristotle distinguished the spheres of the individual and the polis. The city is "built politics", a "legal and moral community of free and equal citizens". The ultimate aim of the polis is not just offering citizens a collective life, but a worthy life. Difference should not just be tolerated but respected as an essential richness of human development. Vitruvius stated that cities should be "solid, beautiful and useful". Alcaeus had already suggested, in the seventh century BC, that "cities are not made from their roofs, stone walls, bridges and canals but from people able to grasp opportunities". This last definition is reminiscent of the definition by Geddes: "The city is a dramatic action".

Population density is the defining characteristic of urban settlements and implies the geographical concentration of human, social, built and economic capital. The proximity of people and activities is a major source of advantages. Cities are the only places where people and resources congregate at a point beyond which synergetic effects become more important than simply accumulative ones.

Each city has a critical mass and a self-engendering and fulfilling capacity. Doxiadis (1974, 1975a, b) characterised the city as the place where different individuals can come together and achieve significant change. If a small village has hundreds of red dots and one blue dot, a metaphor for the one notably different person (e.g. a wise person, a saint or a fool) distinctive from the population symbolised by the red dots, as the settlement increases in size, the number of blue dots also increases. As the scale grows, blue dots may come together, reinforce each other and impact on the surrounding red dots, provoking an overall change.

The cities of the third millennium seem to be more heterogeneous, multicultural, multiethnic and multiple than ever. The challenges became greater and even planetary and require more holistic, multifaceted and synergistic approaches, beyond processes linked to linear developments. Pyramidal relationships have been enriched with horizontal networks. Cities are shared experiences, made up of sights and scents, relations and conflicts, of convergences and differences, of facts and legends, places where lifestyles are shaped. Seedbeds of innovations and bastions of culture, cities are more conscious of the power of the digital economy to extend their endless frontiers and optimise their concentration of knowledge and information (The Economist 2007; Ekistics 2002).

1.1.2   Cities and Demography: The Fundamental Facts

Ancient Athens introduced the world to the interrelated concepts of polis and democracy. Rome became the politically most important, richest and largest city for almost 1,000 years. From one million inhabitants by the end of the first century AD, it was reduced to a ruined and dispersed settlement of 20,000 by the early Middle Ages. Baghdad reached a million inhabitants around 775 AD, shortly after its foundation, and Angkor, once capital of the Khmer empire which flourished between the ninth and the fifteenth centuries, could have sheltered over one million people. Beijing reached this size in 1850. Asia is again the continent of megacities after an eclipse of 150 years which saw the emergence of new agglomerations starting from the developed world and stellar growth in developing countries.

Each country has a portfolio of settlements of different sizes. Each agglomeration is a balancing act between centripetal and centrifugal forces, the equilibrium point depending on the economic activities and the sociocultural and political context. The urban population is the one de facto living in areas classified as urban according to the criteria used by each area or country. The administrative reclassification of some settlements as cities in countries such as China and India could change significantly the estimates of the world urban population.

Human settlements are the vital points in the nerve systems of countries. Although the growth of cites seems anarchic, a country's urban hierarchy is characterised by two robust regularities. Zipf's "rank-size rule" stipulates that the national rank of a city and its population are related in a linear way, whereas Gibrat's law highlights that a city's rate of population growth tends to be independent of its size.

The evolution of cities follows cyclic lines that constitute the phenomenon of "urban transitions". These urban waves are linked to the geographical position of cities, the architecture of the national territories and the nature and pace of economic growth (World Bank 2009).

The first transition involves the movement from a primarily agrarian economy to an industrial economy. It is characterised by urbanisation and the fastest growth in the city centre. The second transition takes place at a higher level of development and involves suburbanisation, with the fastest growth at the suburban ring. The next stage is linked to the transformation into a service-oriented economy and is characterised by counterurbanisation, marked by a population decline, mainly in the urban core and often even in the surrounding suburbs and exurbs. This characteristic is also called "rurbanisation", as rural regions benefit from the loss of the urban population. Last but not least, with reurbanisation the core grows again faster than the suburbs. Most developed cities are at the third or the fourth stage of the cycle. Many of them have even initiated a new cycle with citizens returning to regenerated city centres or even indicating symptoms of overall dissolution of the previous patterns.

The world has been urbanising for centuries. Industrialised regions led urbanisation and reached the "urban age" half a century ago. The gap in urbanisation levels increased between 1950 and 1975 and has been narrowing since 1975. Urban dwellers represented already 50% of the total population in OECD regions in 1950 and 70% in 1975. Waves of intense urbanisation have followed periods of economic growth in northern Europe and subsequently in the USA, Japan, Australia and New Zealand.

The world annual rate of urbanisation has been estimated to be 0.83% during the first 5 years of the millennium and 0.82% during the years 2005–2010. The UN World Urbanisation Prospects database, providing population data for 229 countries since 1950, is a wonderful source of information, but the data are based on national definitions and have often to undergo significant adjustments. The share of the urban population in the total population growth in the less developed regions exceeded for the first time that in the more developed regions towards the end of the 1970s. Since then, most urban growth has taken place in developing countries, fuelled by both rural–urban migration and natural population increase (UN 2004, 2007).

The urban growth rate has been declining since 1980 in both the more developed and the less developed regions. In spite of this decrease, the average absolute annual increment is steadily becoming larger. Paralleling the shift of the demographic centre of gravity towards the less developed regions, more than 90% of the global urban increment comes from the less developed regions. Within the range of less developed regions, the least developed countries have both a lower rate of urbanisation and faster urban growth.

Europe's share of the world urban population is decreasing. In 1950, about two thirds of the urban population in the more developed regions resided in Europe. The urban population growth rate reached its peak (1.05%) during the period 1960–1965 and then declined to 0.16. The share of urban population was already more than 60% in 1970.

In 1999, the world population reached six billion. It is expected to be seven billion before 2015 and eight billion by 2025. It was one billion in 1804, two billion in 1927, three billion in 1960, four billion in 1974 and five billion in 1987. The UN World Population Prospects indicate that, presently, the population growth rates are falling faster, fertility declines are broader and deeper and migration flows are become larger. More than 80% of the world population live in less developed regions and more developed countries are losing their relative population weight. Asia, with an estimated 4.2 billion in 2010 and a population density of 131 inhabitants per square kilometre, accounts for 60% of the world population, Africa for 14%, Europe for 10.6% and North America for 5% (UN 2009).

The process of economic urban development is closely linked to the phenomenon of demographic transitions. The causal mechanisms of a demographic transition include economic growth, education, health care, rural exodus, family structures and employment patterns, especially linked to female labour and public policies. The forms and speed of transitions vary greatly in space and over time, but share some broad characteristics.

Population growth was slow until the Industrial Revolution, with all countries experiencing high birth and death rates. In the era of industrialisation, and as many world countries reached a high level of socioeconomic development, death rates fell drastically, due to breakthroughs in science, better conditions and improved health care and hygiene. At the same time, birth rates remained high and increased after wars, resulting in a very rapid population increase.

Since the early 1970s, a second transition has taken place in OECD countries, resulting from declining birth rates. As birth rates fall, the population stabilises at a new higher level. This transition is taking place in many developing countries, but stabilisation in the population could take decades, since these countries have very young populations and a large number of females in the birth-giving ages. The second demographic transition has resulted in ageing population in many countries.

The world population growth rate of approximately 1.18% per year is the lowest since the Second World War, significantly less than the peak growth rate of 2.02% in 1965–1970. Average annual population increments exhibit a time lag of about two decades. The year 1990 marked a historic turnaround. The absolute population increments reached their highest of 88 million persons per year added to the world population and have been decreasing since then to around 79 million estimated for 2010.

The world average figures conceal large differences across countries and regions. The population of the less developed regions increased by 1.42% during the first decade of the millennium, whereas the population of the more developed countries increased by only 0.35%. It is also important to highlight the increasing demographic weight of the 48 least developed countries, which have a population growth rate of 2.33%. They experience higher fertility, higher mortality and urban population growth higher than their population growth.

Europe is experiencing a low population growth rate (0.08–0.09%) and the natural increase is starting to be negative in several countries. Most of the population growth in the EU is due to immigration gains. Since 1990, central and eastern Europe have experienced population decreases, due to emigration, sharp fertility declines and rising or stagnant mortality. Within the EU, after years of steady decline, the total fertility indicator reached the bottom of 1.42 births per woman in 1995, but an upturn has been observed since then (UN 2009).

The world average fertility stands at 2.5 births per woman, half the fertility rate of around five births per woman in 1950. Birth rates are declining in all regions of the world. Africa has the highest fertility rate of 4.7 children per woman. In most OECD countries, fertility is significantly below 2.1 births per woman, which is the fertility rate necessary for the replacement of generations. The USA and Mexico exhibited, during the years 2005–2010, fertility rates of 2.09 and 2.3, respectively, much higher than Europe (1.50), which has a net reproduction rate of only 0.71. In the Nordic countries, in particular Sweden and Norway, fertility substantially increased in the early 1990s before declining again (UN 2009).

Global life expectancy at birth is continuing to increase, although with high differences across the globe. The estimated life expectancy at birth in the more developed regions (77 years) is more than 10 years higher than in the less developed regions (66), which, in turn, is more than 10 years higher than in the least developed regions (56 years). Japan has the highest life expectancy in the world (82.7 years), followed by Iceland and Canada. Among the world regions, North America and Australia exhibit the highest life expectancy rates. At the other end,

Africa has the lowest average life expectancy at birth (54 years) and the highest infant mortality, with 123 deaths of children under 5 years of age per 1,000 births against eight deaths of children under 5 years of age in more developed countries. Gender differentials in life expectancy, favouring women, by 5 years on average, exist in most world regions. They tend to increase when overall mortality levels decrease and they are highest in Europe (8 years).

The global health situation is improving but new risks are emerging. A third of the world population is undernourished and almost a third of the population of the planet has a diet which suffers from various qualitative deficiencies. The long-term factors which influence the cost of food at world level (demand, weather, fuel consumption, financial speculation, stocks and energy and fertiliser prices) could lead to future increases. On the other hand, more than a third of the population of the USA is obese. The number of the world citizens suffering from diabetes, primarily type 2 diabetes, increased from 30 million in 1985 to 246 million in 2005.

The increasing mobility, interdependence and interconnection of the world create a multiplicity of conditions favourable to the rapid propagation of infectious diseases and of radionuclear and toxicological threats. The expansion of megacities in many developing countries could contribute to amplifying the risks of propagation of diseases and of health deterioration.

Pandemics generate high global fears, especially among highly concentrated urban populations. In its third decade, the acquired immune deficiency syndrome (AIDS) epidemic is still a cruel demographic plague. In 2006, a political declaration on HIV/AIDS adopted unanimously by the UN member States provided a strong mandate to help move the AIDS response forward, with scaling up towards universal access to HIV prevention, treatment, care and support. In 2007, 33 million people were living with HIV. UNAIDS (2008) revealed that the global percentage of people living with HIV has stabilised since 2001, but at an unacceptably high level.

In many countries, the spread of AIDS led to a dramatic drop in life expectancy and erased decades of progress in improving living conditions. The impact of HIV/AIDS is particularly devastating in sub-Saharan Africa. In the hardest-hit African countries, the average life expectancy at birth is currently almost 10 years less than it would have been in the absence of AIDS. In Botswana, life expectancy at birth, estimated at 65.7 years during the period 1990–1995, fell to 48.7 during 2000–2010 (UN 2009).

Population change is also due to international migration, which has gained importance during the last 30 years. Cities are the main destination magnets for the world migrants. With increasing globalisation, world population movements became more significant. The world stock of international migrants increased from 75 million in 1965 to 175 million in 2000, one out of every 35 world citizens.

The unequal global distribution of resources and capabilities is a major driver for migration. International migration is expected to continue and to help balance labour supply and demand across the globe. Without an important inflow of immigrants, the European population would start to decrease from 2012. Migration may also be driven by political instability, as in the case of refugees and asylum seekers. Populations also move for ecological reasons, related, for example, to natural disasters due to climate change. There are expected to be 150 million "climate refugees" in 2050.

Immigration is higher in more developed countries, where it has been estimated to be at the level of 2.2% of the population during the years 2005–2010. Europe and the USA had the highest levels of international influx migration, 1.8 and 3.3%, respectively during the years 2008–2010. Migration can expand horizons and choices, but the opportunities offered to the ones who are best endowed are much higher than those offered to the ones with limited skills and assets. Public policy could help correct these distortions and cities could have an important role in developing a welcoming atmosphere for migrants (UNDP 2009).

The population of the more developed world regions is expected to reach a peak of 1.28 billion in 2035 and is then expected to decline. In the 2030s, the population of the currently least developed countries will be larger than the population of the more developed regions. Until the year 2020, the highest growth rates will continue to be in sub-Saharan Africa, the Middle East and North Africa. This pattern will, however, continue to differ from the pattern of the absolute increase in population. The largest increments are expected in South Asia, East Asia and sub-Saharan Africa.

Population momentum is estimated to be the most important contributor to population increase. Even though in many countries fertility rates have fallen below the replacement level, the population will continue to grow as large cohorts move through the reproductive ages, generating more births than are offset by death. Most OECD countries have population momentum between 1.0 and 1.6; however, Japan, Finland and Germany seem to be going towards population decline. The main increase of the OECD population is expected to occur in the USA (OECD 2008a).

The population density in the more developed countries (23 inhabitants per square kilometre in 2010) is less than half the median world population density, estimated at 51 inhabitants per square kilometre in 2010. However, the average population density hides huge differences. The density patterns of Australia, followed by those of North America, on one hand, and the patterns of Japan and Korea, on the other, occupy the two ends of the density spectrum. Europe lies in the centre (32 inhabitants per square kilometre in 2010), with Scandinavian density patterns overlapping with those of North American and the Dutch patterns ranging between those of Korea and Japan. Europe has the densest world network of cities, with an average distance of 16 km between urban areas, against 29 km in Asia and 53 km in North America.

Age distribution highlights the rapid ageing of the world population owing to a decreasing fertility rate and increased longevity. The decline in the number of children under 15 years old is worrying. At the other end of the spectrum, the proportion of elderly, aged 60 or over, is increasing. The "older old" (*fourth age*) are becoming more numerous. Approximately 1.5% of the world population is over 80 years old and in OECD countries, the proportion is higher than 4%. The world median age, dividing the population into two equal halves, increased from 24 years in 1950 to an estimated 29.1 years in 2010. Europe has a median population age estimated at 40.2 years, whereas that estimated in less developed countries is 19.9 years and in least developed countries is 26.8 years (UN 2009).

The industrialised world has been leading the process of population ageing. It is experiencing an unprecedented increase in the number of the elderly people as a proportion of the total population. The population over 65 years in the more developed countries, which was 7.9% of the total population in 1950, reached 10.8% in 1975 and it is estimated to be 15.9% in 2010. Over the next 20 years around 70 million people will retire, to be replaced by just five million new workers. This contrasts strongly with the past generation and the replacement in the workforce of 45 million new pensioners by 120 million "baby boomers".

Europe is experiencing very high levels of ageing. In Italy, the percentage of the population over 80 years (estimated at 6% in 2010) has already overcome the population under 5 years (estimated at 4.8%). The world age dependency (dependent population as a proportion of the working age population) is increasing and this has crucial implications for social security systems. The ratio of retired to working population is expected to double in Europe by the middle of the century and reach 50%.

Unbalanced age pyramids may have very important effects on the ability of societies to provide the labour force that is needed to produce the expected economic growth, provide social welfare, enhance environmental quality and support sustainable development. Selected migration can help balance age pyramids and social welfare systems.

1.1.3 Scanning the Horizon of the Future

Path-finding analyses have highlighted that, in 2025, the world economy will almost double, whereas the population will increase by about 25%. The group made up of China, India and Korea will have as much weight as the EU. The population of India will approach the population of China, which will have started to decrease (EC 2009e).

Concerning the future distribution of the world population, the UN's low, medium and high variants of population projections based on past trends and possible future developments regarding fertility, mortality and migration yield similar results. The shares of the Asian and Latin American populations as a proportion of the world population are relatively stable at approximately 60 and 10%, respectively, whereas a 1% decrease per decade of the share of European and North American populations is expected to compensate for the increase of the African population (UN 2009).

In 2025, the population of the EU will account for only 6.5% of the world population. None of the countries of the EU taken individually will count among the ten most populated countries of the world. The EU will have a very high proportion of people of more than 65 years old in the world, 30% of the total population. In 2030, just over two people per elderly person are expected to be working age, compared with four in 2008.

The USA–EU–Japan triad will no longer dominate the world, but global leadership will be more distributed. The emerging and developing countries, which accounted for 20% of the world's wealth in 2005, will account for 34% of it in 2025. China could become the second world economic power from 2017. Around 2023, India would become the sixth economic power of the world, ahead of Italy and behind France. In 2030, the "global middle class", with an income between 4,000 and 17,000 dollars a year, could account for one billion people, with 90% of them living in developing countries.

Concerning world trade, in 2025, the volume of trade may have doubled, in relation to 2005, with an increasing part of exports coming from the countries of the South (34% compared with 20% in 2005). The positions of Asia and the EU are likely to be reversed. The EU is not expected to be the leading world exporter any longer. The exports of the EU (39% of the world volume in 2005) are expected to account for only 32%, whereas the share of Asia could increase from 29 to 35% (EC 2009e).

The distribution of natural resources on Earth is very uneven and very different from the distribution of wealth. More than 50% of the major ore reserves are located in very poor countries. For certain metals of high technological importance, the EU is very dependent on imports and access to these raw materials is becoming increasingly difficult. More than 50% of the major ore reserves are located in countries having a per capita income of US $10 or less per day. Many countries that are rich in resources apply protectionist measures which stop or slow down exports of raw materials to support their downstream industries. Half of the growth of the consumption of these products from 2002 to 2005 was due to China, which increased its investments in the mining industries of African countries.

The progress achieved during the last 20 years in the fight against climate change has remained limited in comparison with what a successful transition towards sustainable development would a priori suppose. Measurements taken by scientists since 2000 have shown that the world emissions of carbon dioxide are growing more rapidly than the most pessimistic scenario. A warming of less than 2 C in a century seems quite unlikely, but much depends on global climate politics and the ongoing negotiations for a post-Kyoto agreement (EC 2009e).

Europe is ready to enter into binding and quantified commitments, unilaterally, 20% reduction of emissions between now and 2020 in relation to 1990, and 30% in the event of a global agreement. Furthermore, Europe proposes long-term ambitious objectives, an overall 50% reduction of emissions by 2050 and reductions of between 60 and 80% for the most developed countries. Climate change probably offers one of the best chances to reframe international relations.

1.1.4 The "Silent Crisis" of Poverty

Since the waning years of the twentieth century, urban growth has occurred almost exclusively in developing countries and at an unprecedented scale and pace. This has resulted in the accelerated urbanisation of both poverty and environmental degradation. Poverty in all its dimensions remains humanity's most serious problem. The urban poor experience the worst impact of air pollution and the poorest housing conditions. The most severe harmful impacts of the climate crisis are also expected in the developing world, further aggravating poverty and socioeconomic disparities.

The "silent crisis" of poverty undermines progress in virtually every aspect of human development. It is not just a lack of income but also a denial of possibilities. Eradication of poverty represents the major universal goal for the twenty-first century. It may reduce the capacity for full enjoyment of benefits and opportunities and breed discontent and anger. Cities have a key role to play since they host most of the world's 2.8 billion people struggling to live on less than US $2 per day.

Human development is uneven and it is of the utmost importance to extend the wealth of possibilities to all world citizens. The UN Millennium Development Goals constitute a compact among nations to fight human poverty. Halving the proportion of people in extreme poverty by 2015 will require a dramatically accelerated pace of progress. Environmental sustainability is one of the eight goals, together with universal primary education, gender equality and maternal health, fighting HIV/AIDS, malaria and other diseases, a global partnership for development and the eradication of extreme poverty and hunger (UN 2005).

Global concerted action is considered to be a sine qua non condition for the reduction of poverty. The 2002 Monterrey consensus reminded the world that development is a mutual responsibility (Sachs 2008; UN 2002). Voices have been multiplied for an inclusive and equitable globalisation which offers to all the possibility to live a fulfilling life (OECD 2007b).

The right to the city has evolved internationally as a concept enabling all inhabitants to access and benefit from opportunities offered by the cities. Government at all levels have to create the enabling environments. Data collected by UN-HABITAT in 70 different cities and 70 countries suggest that the right in cities, as social equalisers, is gaining ground. They have revealed that the poor are excluded to a level that prevents the realisation of certain rights and this may have a dramatic impact on sustainable development and access to the world of knowledge and ideas (Brown et al. 2008).

1.1.5 Megapolises, Metropolises, Megacities: Powerful yet Vulnerable

The megalopolises or metropolises, "matrices of civilisation" and huge networks of infrastructures and social interactions, face special challenges. They usually have higher densities as they grow, shorter distances as economic actors come closer and greater diversity of people, patterns and lifestyles. They shelter an increasing proportion of the global urban population, even though more than half of the world's urban dwellers continue to live in cities with less than 500,000 inhabitants. Twenty-three percent of urban dwellers live in a city with a population between one million and five million, whereas 9% of the world urban dwellers live in megacities with more than ten million inhabitants.

The growth and multiplication of giant urban agglomerations, dominating the surrounding areas and, increasingly, the world, has been one of the inheritances of the twentieth century. Even at very diverse stages of development, metropolises share common concerns on a broad range of issues, from their physical infrastructures, to the conditions for the information society and the digital economy and the search for performing models of metropolitan governance (Metropolis 1996, 1999).

In 1950, the world had only two megacities, defined as metropolitan areas with more than ten million inhabitants, both in developed countries (New York and Tokyo). Since 1970, the Tokyo agglomeration, with a population of 35 million in 2005, is by far the most populous urban agglomeration in the world. It dethroned New York, which ranked first in 1950 and 1960, and is expected to be dethroned by Mumbai in the next decade. Mexico City, Seoul, São Paulo, Kolkata, Lagos and Manila are on top of the pantheon of megacities, all belonging, with the exception of Lagos and Manila, to G20 countries (UN 2007).

The developing world is the most represented grouping in the list of the top world largest agglomerations. India is the only country with three urban agglomerations on this list (Mumbai, Kolkata and Delhi). In contrast, European cities are gradually disappearing from the list of the most populated agglomerations. It is expected that by 2020 the majority of the world megacities will be located in Asia.

The expansion of some developing large cities has been explosive and unbridled. Many megacities in the less developed regions exhibited a growth rate exceeding 7% and even 10%. In Bangladesh, Dhaka, the most densely populated city in the world, doubled its population between the 1990 and 2005. In Tanzania, the population of Dar es Salaam doubled within 12 years. In contrast, the megacities of the developed world experienced low annual growth rates during the two last decades, less than 1% for New York and Osaka and just over 1.5% for Los Angeles and Tokyo. Mexico City, together with the two Latin American megacities (São Paulo and Buenos Aires) and the two Chinese ones (Beijing and Shanghai) grew by less than 2% on average during the same period.

The world megalopolises, which host one billion world citizens living in slums, informal and illegal settlements, are huge reservoirs of untapped potential. They are characterised by rapid urbanisation, overcrowded and poorly serviced housing, inadequate infrastructures for water supply and sewage collection, disposal and treatment, and lack the financial resources and governance structures that underpin development, democracy and political stability. They often suffer from international overconnectedness and local disconnectedness. In many of the less developed megacities, the supply of basic infrastructure falls far short of demand and health problems, mainly linked to water contamination, affect large parts of the burgeoning population. The cost of quality water is very high compared with income and there is urgency to develop the proper institutional and financing mechanisms.

Metropolises are dense and intense spaces and societies, which can be very influential. They do not only dominate the world because of their sheer size, but they also produce models which are quickly disseminated to the rest of the planet. They develop powerful networks and interactions among them and cities of every size and nature and enter world markets to take advantage of scale and trade in specialised products. Joint green public procurement of smarter and leaner products could have a colossal effect on the consumption patterns and behaviour economics of the planet (The Megacities Foundation 2008).

Demographic growth, social and economic development and environmental conditions are increasingly being seen as closely linked together. The UN Conference on Environment and Development (Rio de Janeiro, 1992) marked a breakthrough in the discussion of population and environment issues. Chapter 5 of Agenda 21, the agreed compass into the twenty-first century, focuses on demographic development and sustainability. The Population Summit (Cairo 1994) and the Social Summit (Copenhagen 1995) reflected the growing international awareness of the linkages and interrelation between urban growth, social and economic development and environmental and natural resources.

Population growth is exacerbating poverty and is a major contributor to increased pressure on the global environment. Ten years after the Cairo Summit on Population and Development, the UNFPA examined the progress made and the obstacles encountered at the halfway point in implementing the action plan. A global survey indicated that developing countries have taken important steps to carry out the recommendations, but progress has been uneven. More than 350 million couples still lack access to a full range of family planning services and more than 500,000 women die every year from pregnancy-related complications. These deaths could be prevented by expanding access to attended delivery and emergency obstetric care. Cities could promote efforts by countries and integrate reproductive health services into primary health care, improve facilities and training (UNFPA 2004).

1.2 The Battle for Sustainability Will Be Won or Lost in (Because of) Cities

The concept of sustainable development was one of the most emblematic terms at the turn of the millennium (UN 2005; UNESCO 2004). Defined as a process and not as an end point, as an itinerary rather than a destination, it may be construed as a journey to Ithaca, as a struggle to resist the song of the sirens of excessive consumption, the Charybdis of economic crisis and the Schyla of social exclusion and environmental degradation. Like Ulysses, "who …saw many cities and learnt their mind… endured many troubles and hardships…", cities have to follow the tortuous path of rationalised patterns and lifestyles to resist tempests and wild winds and invest the lessons of the journey in a better world.

Cities seem decided to lead the world towards sustainable development. The first reason is simply their sheer size and the extraordinary concentration of resources, people and activities. Although average world figures conceal large differences across countries and regions and should be considered with caution, cities host more than 50% of the inhabitants of the planet but consume around 75% of its resources and produce 75% of the overall greenhouse gas emissions. Responsible for the great majority of waste, pollution and emissions, cities are also responsible for addressing their prevention and reduction and for leading adaptation processes.

Compared with the average world citizen, each urban dweller consumes fewer resources and is responsible for lower levels of emissions than the rest of the inhabitants of the planet. Resource and energy efficiency is higher in cities and can be increased at a lower cost owing to large scale economies. Action in cities represents the most efficient way to address the global challenges.

The cultural reasons are equally important. Cities have the capacity to engender new values, create new concepts and introduce structures and patterns, which are quickly disseminated to the rest of the world. They are beehives of creativity and seedbeds of innovation. Because of such powerful urban cultural, economic and political influences, cities are the ideal geopolitical medium for sustainability-related projects and awareness campaigns.

There are also fundamental political reasons. Cities have always promoted local democracy and this is a major precondition for advancing towards sustainable development. Cities bear witness to the will not just to adapt to change, but also to lead and drive the transition to sustainable development. An extraordinary number of symbolic events, awards and projects reveal a mine of ideas and energy invested in action.

The most well known definition of sustainability, after the historical definition proposed in the Brundtland report, declaring the obligation to future generations, is the three-pillar paradigm for the temple of sustainable development supported by economic, social and environmental columns. The definition has been criticised for not highlighting the political dimension, but one can suggest that this dimension and also cultural aspects are included in the social pillar. Sustainable development advocates a new dynamic balance between quality and quantity of development and asks for optimal integration of socioeconomic and environmental concerns supported by active citizen participation (EC 2003a, 2005b, 2007f; EFILWC 1997a–g; Hall and Pfeiffer 2000; Mega 2005, 2008; WHO/OECD 1996).

Sustainable cities could be defined, according to the scientific literature, as the cities which do not cause the consumption of renewable resources at a rate higher than their regenerative capacity and contribute to the replacement of non-renewable resources, particularly through savings, research and innovation. Some suggest that the lofty promises of the sustainable cities may constitute an oxymoron, as they are almost impossible to realise given their high dependence on resources imported from other places (EC 2009a).

"Making the leap greater than the decline" (Odysseus Elytis) could be a most poetic definition for sustainability. Many cities strongly advocate halting the depletion of natural resources and preserving and enhancing the urban capital, with its tangible and intangible aspects. A sustainable city is, first and foremost, a city which carries in it the seeds of progress and preserves its capacity to reinvent itself and grasp the opportunities of the future.

The move to the civilisation of sustainable development demands further and faster progress against the depletion of the capital stock, composed of human and social, natural and environmental resource capital and human-made capital. The human capital includes personal skills, ethical values and lifestyles, whereas social capital comprises cultural patterns, traditions, networks and institutions. The environmental resource capital includes the life support systems (land, water, air), offering their services for free, whereas the natural capital comprises the physical endowments (built and open spaces, public spaces), assets typically transacted in markets. Progress can be estimated through the evaluation of the increase or decrease of the aggregate capital stock.

A key question concerns the extent to which different components of wealth can be substituted for each other. Strong sustainability implies that substitution of social and environmental capital is impossible and loss is irreversible. A minimum of environmental, social and economic capital forms a triple bottom line. For weak sustainability, the matter is not whether a particular natural resource will be available infinitely, but whether human ingenuity can preserve and increase the global capital (OECD 2001a).

To overcome the present crisis, advancing on one pillar of sustainable development critically depends on improved performance on the other two pillars. And, on the other hand, weaknesses on one pillar undermines progress on the other two. Policies have to incorporate strong sustainability principles for advancing simultaneously and harmoniously on all pillars. Environmental improvement cannot be disconnected from economic progress and social justice.

The 2008 crisis demonstrated that, in the long term, economic excellence risks being undermined by social and environmental failures. Poverty and the irretrievable loss of ecosystems and biodiversity have been fatal to the US economy, the first economy of the planet. The USA is the OECD country with the highest inequality and the highest poverty rate after Turkey and Mexico and equity is at the heart of sustainable development.

Sustainability requires a high level of political commitment to have the three pillars working together, a long-term vision and a willingness to open up the decision-making processes to all stakeholders. Each city has to strike a fine dynamic balance among economic, environmental and sociocultural development goals, in constant coevolution, underpinned by active citizen participation.

The preparation of Local Agenda 21 plans was the epitome of the concept "think globally, act locally." The Local Agenda 21 process includes the management and improvement of the environmental performance and the integration of sustainable development aims into all policies and activities. Following the review of 10 years of Local Agenda 21 plans, implemented by 6,400 local governments in 113 countries, at the Johannesburg summit, the Local Government Session provided the context for the launch of Local Action 21. Local Action 21 incited moving from agenda to action, from plan to practice.

In reaffirming the Rio principles, Local Action 21 became simultaneously a motto for accelerated implementation of sustainable development, a mandate given to local authorities worldwide and a movement of most active cities towards sustainability. The "Johannesburg call" urged cities to be bold and unequivocal about changes and to develop very realistic and pragmatic action plans.

The preparation of local agendas created a global momentum for the enhancement of the urban environment. Chapter 28 of Agenda 21 specifically addresses local government action and asks for the most active involvement of governments, local authorities and citizens. It is a major commitment for any local authority, which ultimately has the responsibility to act not only on behalf of its present citizens but also its future citizens. Institutional fragmentation is sometimes a barrier to holistic approaches. Local governments have to commit to the local environment and economy, equity and social development. They should become more open, flexible, proactive and responsive.

At the international horizon, the leadership of cities on sustainable development has been progressively strong and engaging. In 1996, the last UN Summit of the second millennium, HABITAT II, preceded by the first World Assembly of Cities, offered governments and cities the opportunity to join visions and actions for urban sustainability. The HABITAT II Agenda focused on the principles of equality, eradication of poverty, sustainable development, liveability and diversity, family, civic engagement and government responsibility, partnerships, solidarity and international cooperation and coordination. Commitments aimed at adequate shelter for all, sustainable human settlements, financing and progress evaluation.

Progress with the implementation of the Global Action Plan, including provisions for adequate shelter, sustainable human settlements, capacity-building and institutional development, international cooperation, coordination, implementation and monitoring, has been slow. Dissemination of best practices and sustainability indicators, suggested as two important instruments for a qualitative leap forward (UN/HABITAT 1996; UN 1997), led to an intense activity in compiling good practice guides and building indicator frameworks. The World Summit on Sustainable Development (Johannesburg 2002) provided a new impetus for a global deal to accelerate progress.

Next to the world summits, the World Urban Forum, a biennial gathering established by the UN in 2002, examines rapid urbanisation and its impact on communities, cities, economies, climate change and policies. Attended by a wide range of urban actors, professionals, academics, officials from governments and local authorities and national and international associations of local governments, it offers a common platform to discuss burning urban issues in formal and informal ways and to come up with action-oriented proposals to create sustainable cities. Since the first meeting in Nairobi, in 2002, the forum has grown as it travelled to Barcelona in 2004, Vancouver in 2006 and Nanjing in 2008. The fifth forum, to be held in Rio de Janeiro, will focus on bringing the urban divide.

The first Monday in October every year, designated by the UNs as World Habitat Day, sheds light on the state of cities and the basic right to adequate shelter. It also reminds the world of its collective responsibility for the future of the human habitat and the importance of the recognition of exceptional action. World Habitat Day is also the day of the attribution of each year's Habitat Scroll of Honour, the most prestigious human settlements award. Launched by the UN Human Settlements Programme in 1989, it already acknowledged initiatives which have made outstanding contributions in fields such as shelter provision, developing and improving the human settlements and the quality of urban life, care for the homeless and leadership in postconflict reconstruction. The city of Malmö was one of the Habitat Scroll of Honour award winners for 2009, primed for its innovative, holistic approach to becoming a twenty-first-century eco-city.

1.3 The Sustainability Dynamics of the European Urban Archipelago

Europe is first and foremost urban, an archipelago of 3,500 agglomerations with more than 10,000 inhabitants, 365 agglomerations with 100,000 inhabitants and 32 agglomerations with more than one million inhabitants. The constellation of European urban systems appears rather stable in terms of demographic growth. The almost zero population growth and the increase in the average age of the Europeans being the main demographic trends in the EU, cities seem more engaged in qualitative rather than quantitative leaps.

History played an important role in the development of European cities. Many of them have even mythological origins and are beacons of millenary civilisations. Large cities seem to have reached their greatest size, the overtaking of which could entail an unbearable burden of environmental problems and social shock waves. The myriad of changes in their microcosm, but also in the macrocosm, affect them in very different ways.

The start of the millennium found European cities competing more among themselves, but also cooperating more. They share their experiences and exchange their green and grey agendas in a noble emulation to win the common battle for sustainable development and to become more attractive to people and capital. Addressing climate change is high on their political priorities. Many of their lessons can be shared with world cities (Beatley 2000).

Europe is a continent of small and medium-sized cities. Only London and probably Paris could claim the attribute "world city." Urban clusters, like the Randstadt (area including Rotterdam, Amsterdam, The Hague and Utrecht), develop in many European countries. Networking is a must for cities willing to build upon each other's experiences. "Small large" cities having an intermediate position in the urban system may play a capital role for the dynamism of the urban network. Medium-sized cities are usually on a more human scale and they offer a better physical and social environment than large cities. Their advantages make them attractive for investments when more powerful cities have environmental and social problems.

Intermediate medium-sized cities are often more open to the countryside and may act as a filter between larger cities and regions. Medium-sized cities on the periphery of metropolitan areas face special opportunities and threats. Their geographical position may provide them advantages of access and networking, but also disadvantages in developing their autonomous identities and their complementary functions (EFILWC 1997d; Bellet et al. 1998).

Since the 1990s, urban issues have been high on the political agenda of the EU. The green paper on the urban environment and the reports on the sustainable city recognised the role of cities as fulcrums of economic activity, innovation and culture. Nearer to the turn of the millennium, the EC Communication "Towards an Urban Agenda in the European Union" expressed the intention to make EU policies more urban-sensitive, whereas the Framework for Action for Sustainable Urban Development insisted upon better coordinated and targeted EU action for urban areas.

The European initiative URBAN, launched in 1994, embraced these policy orientations. Urban I (1994–1999) allowed 118 cities in the EU to improve their living and working conditions, whereas Urban II (2000–2006) assisted 70 sensitive urban districts in their efforts in terms of competitiveness, social integration and physical and environmental regeneration (EC 2003a). The first evaluation of Urban II, presented at the conference "Cities of the Cohesion", held in London in 2002, was followed by the signature of a statement requesting a strengthening of the urban actions at the level of the EU.

Following the Urban initiatives, the Urbact programme confronts and capitalises on the experiences in the framework of the above-mentioned programmes. Urbact promotes exchanges between cities thanks to the creation of thematic networks, conducts analyses to propose innovative action and disseminates the exemplary practices to all European cities. In 2007, 43 cities of the member States which joined the EU since 2004 benefited from the initiative "Urbact Support for Cities", through the visit to cities and the provision of advice by Urbact experts.

At the EU level, the European strategy of sustainable development, adopted in Gothenburg in 2001 and reinvigorated in 2006, includes explicit mentions of the urban problems, including congestion, deterioration of the vital nerves of the cities, urban sprawl, poverty and social exclusion (EC 2001). In 2006, the European Commission adopted a thematic strategy on urban environment, structured around four priorities, selected after consultation with the social actors: the urban environment, transport, sustainable construction and urban design. For each topic, the Commission presented the challenges, the actions already in progress and prospects for future actions (EC 2006a). Furthermore European programmes provide possibilities for training and capacity building for local authorities, to develop the skills necessary for the management of the urban environment.

In the European Union, member States and the Council of Ministers have also been active in urban affairs. In 2004, under the Dutch presidency of the EU, a number of common principles, known as "Rotterdam acquis", were adopted for successful urban policies. The national governments committed themselves to support the urban strategies, including those of small and medium-sized towns, in their efforts to achieve the Lisbon objectives for Europe to become the most dynamic knowledge-based economy in the world. In 2005, the UK presidency underlined the concept of the sustainable communities and the importance of cooperation to acquire the necessary skills. A symposium organised by the Academy for Sustainable Communities in 2006 analysed the question of the strategic competences for the period 2007–2013 (Leeds Metropolitan University 2007).

During the German presidency, in 2007, ministers in charge of urban affairs signed the "Leipzig Charter on Sustainable European Cities" (Leipzig, May 2007), a statement of shared principles for the urban development policy in Europe. The charter recognises European cities as parts of a common heritage. The reinforcement of distressed areas, the employment of young people, the fight against exclusion and sustainable architecture are some of the aspects of the "European city" highlighted by the Leipzig Charter. The recommendations cover all facets of the urban development policy, including creation and preservation of dignified public spaces, modernisation of the infrastructures, better energy efficiency, active innovation, education and training (Council of Ministers 2007).

Moreover, with the territorial agenda, the European governments engaged to promote polycentrism and innovation by cooperation between cities, to create new forms of partnerships between the cities and the countryside, to cement the regional clusters to increase competitiveness, to strengthen the trans-European management of the risks, such as the effects of climate change, and to reinforce the ecological structures and the cultural resources. The action plan for the implementation of the agenda was adopted during the Portuguese presidency of 2007. The Slovenian presidency of 2008 wished to develop synergistic dynamics through the analysis of the results of the coordinated actions for territorial cohesion and urban development.

The conference "Cities, Places of Every Possibility" with the eloquent subtitle "From Lisbon to Gothenburg, Cities Make Europe", during the French presidency of 2008, suggested that the large cities offer privileged ground to reconcile Europe with its citizens. The declaration of the ministers in charge of urban development (Marseilles, November 2008) brought forward the tetraptych of challenges for cities, including environment and climate change, competitiveness, social cohesion and citizenship. It recognised that 69% of the greenhouse gas emissions in the EU are produced in cities, which are also the key actors for reducing them. Finally, the declaration suggested that the urban dimension of the Lisbon and Gothenburg strategies should be reinforced (Council of Ministers 2008).

In the aftermath of the Marseilles ministerial declaration, work was launched to provide a concrete follow-up to the Leipzig Charter on Sustainable European Cities. Two working groups have been established, the Member States/Institutions Group (lead by France) and the Cities Group. The Cities Group, created to serve as a testing group, is running under the Urbact II programme and it is led by Leipzig, with partner cities Rennes, Gothenburg, Bytom, Kirklees, Szekesfehervar and Vitoria-Gasteiz.

In 2009, the Czech presidency organised the European Summit of Regions and Cities, focusing on cohesion policy. The topics also reflected economic developments in the world, including Europe's ability to react to the financial crisis and find local solutions to address global challenges.

Opening themselves and to the world is essential for cities to progress towards sustainable development. Since 2003, the open days of the European Week of the regions and cities, organised each October by the European Commission and the Committee of the Regions, bring together in Brussels thousands of participants from all the regions of Europe. "Coffee of the investors" serves as a platform for the public and private investors. In 2009, the seventh open days attracted 7,500 participants to more 125 seminars and other events in Brussels and 230 additional local events in the member States under the aegis of the open days. They shared ideas on and insights into a wide spectrum of issues, ranging from boosting innovation and counteracting the recession to the long-term challenges of climate change.

On the road from Rio (1992) to Istanbul and HABITAT II (1996), the first conference on European sustainable cities and towns (Aalborg, May 1994) marked a paramount step in the move towards urban sustainability with the signature of the "Charter of European Cities and Towns: Towards Sustainability", the starting point for the European Sustainable Cities and Towns Campaign. This campaign, a massive movement of cities in Europe, is an important pillar in the pantheon of world networks and movements (ICLEI 1995).

The second conference on European sustainable cities and towns (Lisbon 1996) urged cities to move from charter to action (ICLEI 1997), whereas the third conference (Hanover 2000) made the leap to the new millennium with the European Sustainable Cities and Towns Campaign Mayors convention declaring local sustainability as their highest political priority.

The Charter of European Cities and Towns: Towards Sustainability, seen as the European version of Local Agenda 21, states that cities and towns should define their living standards according to the carrying capacity of nature and advance towards social justice, prosperous economies and environmental improvements. Social equity is considered to be a precondition to the achievement of sustainability, as the inequitable distribution of wealth both causes unsustainable behaviour and increases resistance to change. The charter advocates the development of urban sustainability indicators as yardsticks of progress.

The charter embraces an ecosystem approach to urban challenges and declares the responsibility of European cities for many of the global problems. The patterns of division of labour and land use, transport, industry, consumption, leisure and, hence, values and lifestyles are not neutral. Sustainable development cannot be achieved without governments, local communities and citizens rising to meet the goal of leaving the next generations, an at least undiminished, capital. Each city has to find its own unique path towards sustainability. Integrating the principles of the charter in its policies reinforces its vigour and forms a common basis for progress.

Equitable regional interdependencies are given importance by the charter, which recognises that cities cannot export problems into the broader environment or the future. Priority is also given to ecologically sound means of transport and the decrease of enforced mobility. Emphasis is placed on the reduction and stabilisation of greenhouse gases, the enhancement of biodiversity and the preservation of the ecosystems. The signatories of the charter committed themselves to develop Local Agenda 21 plans in partnership with their citizens and asked for sufficient powers and solid finances to carry out projects heralding sustainability (ESCTC 1994).

The European Sustainable Cities and Towns Campaign has been supported by a partnership of important networks and associations of local authorities, which wish "not to have to reinvent the wheel". The Association of Cities and Regions for

Recycling and Sustainable Resource Management (ACR+), Climate Alliance, Council of European Municipalities and Regions (CEMR), Énergie-Cités, Eurocities, International Council for Local Environmental Initiatives (ICLEI), Medcities, Union of the Baltic Cities (UBC), World Health Organisation (WHO) Healthy Cities Project and the World Federation of United Cities (FMCU-UTO) joined forces to promote the European network.

The Sustainable City Award created by the European Sustainable Cities and Towns Campaign aims at recompensing best practice and promoting emulation. The campaign recognised the active, progressive and firm commitment of cities such as Ferrara, Heidelberg and Oslo to the sustainable development process. In these cities, sustainability becomes an innovative, creative and proactive process embedded in the institutional culture and practices, and the subject of a regular, effective and meaningful dialogue with their citizens and their stakeholders.

The Charter of European Cities and Towns celebrated 10 years of visions and efforts with the "Aalborg+10" Conference in 2004. This fourth conference on European sustainable cities and towns was marked by the signature of the Aalborg Commitments, devised to assist cities and towns in achieving sustainability and signed by 110 representatives of local governments. The ten Aalborg Commitments, signed later by more than 600 local governments, is a tool to help local governments to set clear qualitative and quantitative targets for practical work to implement the urban sustainability principles of the Aalborg Charter.

In 2007, the fifth conference on the European sustainable cities and towns (Seville 2007) took stock of the experience of the cities since the adoption of the Aalborg Commitments. The Seville conference brought together more than 1,500 elected representatives and the European cities and reconfirmed, once again, the responsibility of cities for sustainable development and their will to exchange their experiences.

1.4   From Resource-Consuming to Knowledge-Intensive Cities

Sustainability means, first and foremost, sustaining the ability to innovate and progress, a permanent aspiration for a better world. To advance towards sustainability, cities have to tap into their unlimited creativity and continuous invention of new opportunities. Pure evolutionary change, incremental steps and merely adaptive responses, within the established rules and procedures, are not sufficient for advancing towards sustainable pathways. Sustainability requires innovations to break with old patterns and maximise and optimise investments in capital, labour and skills. Setting aspiration levels and targets is an excellent tactic for accelerating the pace of progress.

Cities have often been exploiting new knowledge and driving forward the frontiers of progress. Innovation is "creative destruction", the key to progress (Schumpeter 1976; OECD 1996). It is a non-linear process encompassing multiple interactions between invention an transformation. Creativity, a precondition for invention, thrives on diversity and versatility. Cities constitute large reservoirs of knowledge and creativity. Multidimensional innovative projects are witnesses to the resourceful visions that cities conceive and develop, to meet the interrelated social, economic and environmental challenges. Success is never a certainty, but inaction is a prelude to failure. It is essential to analyse the kernel of the innovations, the definitive spark, which, together with the other elements, brings the desired significant change (EFILWC 1997g).

Knowledge is an inexhaustible resource and the competitive advantage in the learning economy which creates assets by investing in "intellectual assets". Furthermore, unlike natural comparative advantages, innovation can be cultivated and improved. In the era of open innovation, cities have to open up and share innovations, increasingly globalised and distributed. The innovation rainbow can allow new ideas, concepts or products to bring about the desired transformation through a combination of research, design and integrated solutions and services than span business and technology.

Innovation is a composite concept which comprises an organisational and institutional response to the opportunities offered by research and technology. Its main drivers are necessity and choice. Innovations originate from scarcity, pure accident, defence, crisis, creative conflict and strategy. Complex problems that inhibit innovation often create a sharper need for it. Innovation may also be the result of a struggle for survival. The present economic crisis could force cities to take a hard look at reality and generate a plethora of new ideas.

Innovation and sustainability share a desire for eternal youth, which implies a process of permanent renewal. In times of crisis, shifts in the economy may prove swift and lethal for cities, which do not change. Sustainability demands optimisation of urban spaces and times, assets and skills. Cities, as very complex systems, are by definition organisations where many new ideas, concepts and products are created, but where difficulties of implementation also abound. Collaborative innovation is essential to make the most out of the collective talent of a city. It requires vision, strategy and tactics, design of tools and methods, concerted action and communication.

Social innovation can be critical for subverting traditional organisational power structures. Governments at all levels have become more enablers rather than providers, but equally, initiators, upstream partners and drivers of innovation. Innovation is inherently risky and has to be supported and encouraged. Governments can facilitate seed capital and start-up financing such as business angel funds and networks. They can infuse competence and entrepreneurship and raise the standards of transparency, participation and accountability. The social partners can drive innovations, finance unprecedented activities, stimulate other actors, create the social climate for acceptance and help increase the local yield. They all share the responsibility of making a city sustainable.

Institutional innovation is essential for creating the conditions for the transition to sustainable prosperity. It often involves restructuring of governance architectures. The Brundtland report had already noted in 1987 that the challenges presented by population growth, economic development and the need to transform agricultural, energy and industrial systems were highly integrated and interdependent. However, the institutions responsible for these issues tend to be independent, fragmented and working to relatively narrow mandates with closed decision-making processes.

The adaptive intelligence and chronic imitation of innovations may reflect an efficient but non-creative use of skills, capital or technology in cities. Trial is limited in the search for the optimal conditions for systematically importing and transplanting innovations. Rapid mobilisation of creativity and innovation can only take place if there is a permanent environment for nurturing and attracting innovation talent and investing in the peaceful incubation of genuinely new ideas and unproven goods and services. Education and research are critical.

Innovation seeds can come from any individual or collective actor, but the creation of new value usually needs many more actors. Much depends on the maturity and the cohesion of the community, the quality and commitment of human resources and the political will. Governments should give special opportunities to the concepts and ideas proposed by those usually without a voice. Empowering them ultimately empowers the entire society.

The assessment of innovative urban projects has to analyse and compare the situations with and without the given project. Evaluation involves forecasting and comparison and has to weigh various options. In the simplest form, one has to consider the costs of stimulating, generating, designing and implementing a project and compare them with the achieved benefits. Sustainability balance sheets do not only consider costs and benefits in straightforward financial terms (the capital and operating costs, and the money earned in return, suitably discounted) but all contributions to the interrelated aspects of sustainable development. Very often, there are also unintended costs and returns not included in conventional appraisals.

The ideal contribution to strong sustainability would involve multiwin benefits. Their return is in the form of sustainability benefit with all its dimensions, environmental, economic and social. The sustainability balance sheet includes external or social costs which may not enter into the financial profit-and-loss calculus of the decision-makers. All the dimensions of urban sustainability have to be considered to appraise the contribution of innovative projects in reinvigorating the natural, environmental, human, social and human-made capital that form the hardware, software and heartware of a city.

The criteria to estimate the contribution to sustainability may be distinguished into result and process criteria. The former include financial (the strict financial return on social investment and also the level of social capital required when affordability matters), environmental (the contribution to environmental sustainability), social (the contribution to social sustainability) and economic (the contribution to sustained economic growth) factors. Process criteria evaluate whether the path followed in reaching the above-mentioned goals is conducive to sustainability. Capacity building and institutional development are deemed essential, together with the transferability of projects and their contribution to continuous overpassing (EFILWC 1997e).

The very notion of an innovation is very relative and ephemeral and it is difficult to make a unanimously accepted taxonomy of innovations. Strong alliances are needed for key actors to create the space and the conditions for the innovations. The architecture of coalitions is very diverse and challenges general rules. Sharing costs and benefits being the essence of any participatory action, the multiplier dimensions in matters of innovation are impressive. Accountability, transparency and communication are decisive for the social acceptability of innovations (Jacobs 1969).

Innovations are powerful processes that can also be dangerous. Radical innovations may change the status quo and cause serious disruptions. They can lead to non-reversible situations and affect the sociocultural equilibrium of a city. Moderation may be needed to prevent tensions between the imperatives for change and continuity. In collaborative innovations, citizen participation can act as a sociocultural net and business participation as a financial one. To breed and promote meaningful innovation, cities have to be forward and outward looking, participatory and able to guarantee homeostasis, change within stability.

Watercolour 2
Curitiba, the Environmental
Capital of the Southern Hemisphere

Curitiba, the environmental capital
of the southern hemisphere

Chapter 2
For a New Balance Among Nature, Humans and Artefacts in Cities

Cities are dynamic and complex ecosystems, open and vulnerable, the only human ones. By analogy to the biological concentration of species in a mutually supportive environment, humans come together in places that optimise their reciprocal benefits. As sociobiospaces, cities require an extraordinary array of material and labour, they have the most complex metabolisms and produce an equally remarkable array of products, waste and emissions. Their role in the transition to a low/zero- and postcarbon society and economy is crucial. This chapter focuses on key elements of the urban ecosystems and principles and actions inaugurating better consumption and production models.

The battle against global warming, the end result of billions of everyday decisions taken in local environments, will be won or lost in cities. Urban policies should consider all measures to radically reduce the ecological and carbon footprints and the environmental shadow of cities. Resource (including waste) management and quality of land, water and air, as well as noise, are of extreme importance to cities that become laboratories of ecological innovation and provide inspiring examples. From eco-buildings to eco-districts and from eco-cities to the our only planet, actions have to be proportionate to the challenges.

2.1 Improving Urban Consumption: A Vital Priority

To offer citizens the possibility to live fulfilling lives, cities have to offer fresh air and water, healthy food, energy for mobility, electricity and heat and cooling, materials and products, which uphold their capacity to renew and provide resources and services. They require a remarkable array of material and labour inputs and, through most composite processes, produce an equally remarkable array of outputs and outcomes.

The consumption patterns of the waning years of the last century and the often predatory lifestyles were at the heart of the 2008 crisis, characterised by excessive borrowing of both financial and ecological capital. The complex urban metabolisms produce tremendous flows of waste, pollution and emissions. The transition to sustainability demands from cities improvement of their performance and disassociation of economic activity from resource use and pollutant release.

V.P. Mega, *Sustainable Cities for the Third Millennium: The Odyssey of Urban Excellence*, DOI 10.1007/978-1-4419-6037-5_2,
© Springer Science+Business Media, LLC 2010

Cities are the main contributors to, and victims of, global, national and regional environmental problems. At each stage of urbanisation, the environmental conditions in cities dramatically improved. The supply of clean water and food and the rise of standards of living needed radical innovations, which then nurtured mainstream practices. Cities now face a range of challenges, such as urban sprawl, the loss of green spaces (greenfields), the development of abandoned or contaminated land (brownfields), distressed urban areas, gridlocked traffic and overstretched infrastructures, which are not addressed by conventional policies. They result in an accumulation of problems, further aggravated by the economic crisis, which often exceed their capacity to offer the quality of life expected by citizens.

All cities influence the surrounding land and affect a much broader environment than the one with which they directly interaction. In the more developed world, cities and regions consume too many valuable resources and produce too much waste and too many emissions. Their ecological footprints, estimated after the evaluation and aggregation of the biophysical capacity of the land needed to produce the necessary resources and to absorb the waste, often exceed even those of national territories.

The global ecological footprint is estimated to be more than 30% larger than the capacity of Earth to renew the resources consumed. The average ecological footprint in the USA is estimated to 9.4 global hectares per capita, equivalent to about ten football fields. The average footprint among EU countries was estimated at five global hectares per capita in 2005, still well above the 2.1 global hectares needed to prevent debt (EEA 2005). It takes about 1 year and 3 months for Earth to regenerate the resources that the humanity needs and absorb the carbon dioxide (CO_2) produced in a single year. This ecological overshoot ultimately leads to the liquidation of the planet's ecological assets and the depletion of the global natural capital which sustains life on Earth (Ecological Footprint Network et al. 2008).

The 2009 overshoot day was estimated to be 25 September by the Ecological Footprint Network, which also suggests that the global recession barely slowed demand and only delayed the 2009 overshoot by 1 day. For the remaining months of 2009, humanity's needs were covered through depleting resource stocks and accumulating CO_2 in the atmosphere. Many awareness events focused on the need to consume less that the planet can sustainably produce in 1 year.

Three quarters of the human population live in countries that are ecological debtors, consuming more biocapacity than they have within their borders. Cities can influence over 70% of the total ecological footprint. The CO_2 footprint, which accounts for the use of fossil fuels, is almost half the total global footprint, and is its fastest growing component; it has increased by 1,000% since 1961 (Ecological Footprint Network et al. 2008).

Agenda 21 identified unsustainable patterns of production and consumption, particularly in industrialised countries, as a major cause of environmental deterioration. The Johannesburg Plan set out a range of actions that countries should take to influence consumption patterns. Policies to influence production patterns, such as cleaner production and eco-efficiency, are in general better developed and understood than demand-side policies, where progress depends on a huge number of individual consumers and citizen lifestyles.

Stabilising and improving consumption is a major issue and an important objective in all sustainable development strategies. The Charter of European Sustainable Cities and Towns recognises that decreasing consumption levels may be an over-ambitious aim. Consuming better goods instead of more goods could be a device for sustainable consumption (ESCTC 1994).

Urban consumption depends on individual lifestyles, social structures and land-use patterns and policies. An ageing population and decreasing household sizes result in higher consumption pressures. The development of the information society brought important structural changes to urban systems. Urban sprawl is the prime cause of increased consumption of energy, resources, transport and land, thereby raising greenhouse gas emissions and air and noise pollution to levels that often exceed the legal or recommended human safety limits.

Sustainability demands strategic long-term efforts to tackle the underlying causes of unbridled consumption. Cities should respect the geophysical and cultural local limits, mobilise invisible economic and social structures and seek symbiosis with the bioregion. Urban environmental planning for sustainability requires a comprehensive interdisciplinary assessment of urban assets, a natural resource information system and the identification and analysis of the policy distortions and bottlenecks. Prevention should be considered an investment and not an expenditure.

The path to sustainable development is long and tortuous. Leadership towards the low-carbon economy is one of the most ambitious aims of the EU. In 2001, the EU Sustainable Development Strategy asked for the integration of principles of sustainability in all policies and suggested an array of strengthened policy instruments. The reinvigorated EU strategy, adopted in 2005, identified seven key challenges and proposed targets, operational objectives and actions. The 2009 review of the strategy revealed that, despite considerable efforts for sustainable development in major EU policy areas, unsustainable trends persist and the EU has to intensify its efforts. As the example of climate change shows, acting at the earliest possible stage brings more and earlier benefits at lower costs than acting later (EC 2009c).

Among the guiding principles, the precautionary principle tries to strike a balance between the freedom of action and the protection of public health and the environment, which may be irreversibly affected. It is subject to scale of values and ethical debates. In the US, the approach is more focused on significant risks and regulation is based on scientific evaluation. In the EU, the approach aims at building a common basis for assessing, managing and communicating the risks that science is not yet able to determine with sufficient certainty. The acceptable level of risk is defined with extreme caution and implies an eminent political responsibility. Respecting the principle should lead to transparent, coherent, proportionate and non-discriminatory actions and at the same time avoid unwarranted recourse to the principle as a disguised form of protectionism.

Sustainability asks for efficient, effective and equitable policy portfolios with an articulated combination of a number of complementary instruments to maximise and optimise the overall impact. Policy instruments include regulations, economic measures, infrastructure and planning, technology and measures promoting citizen participation. Information and awareness raising are organic parts of these portfolios.

The UNECE Convention on Access to Information, Public Participation in Decision Making and Access to Justice in Environmental Matters (the "Aarhus Concention", 1998) marked an important step in the appreciation of environmental goods.

The absence of markets and property fees for air and ecosystems are at the origin of their unlimited use. It is hard to think of any economic activity that does not benefit from ecosystem services or does not, in some way, impact on ecosystems (WBCSD 2009). Economic instruments, applied through the fiscal infrastructure of a city, act like substitutes for markets that are costly and difficult to organise. Their success depends on the consensual introduction of the instruments, the metapolicy and the institutional context.

Regulatory measures range from the insulation of new buildings to the mandatory installation of water-saving systems. Economic instruments try to ensure that the price of household energy, fuels, water and waste fully reflects the associated environmental and social externalities. They range from differential taxes to encourage the use of environmentally friendly fuels to taxes on municipal water supplies and waste. Social instruments comprise a series of means to raise consumers' awareness of the adoption of more sustainable lifestyles. They range from cartoons and comic strips to eco-festivals and touring exhibitions.

Economic instruments can impinge upon lifestyles and may induce citizens to choose the most efficient and effective energy options in response to the price signal they provide. They promise lower cost of compliance than legislation, reduced bureaucracy, incentives to improve over time, better environmental performance and the generation of funds to finance sustainability policy measures. The use of charges and taxes has to reflect the scarcity value as a powerful tool for sending signals to producers and consumers. Prices impact remarkably on the vitality and ingenuity which markets can unleash. A carbon tax or some other mechanism ensuring that the prices for fossil fuels fully reflect the harm to the environment and public health would be a potent force for decarbonisation. As taxation is unpopular, many governments opted for revenue-neutral tax reforms, which, in parallel to eco-taxes, reduce the tax burden on labour. Such reforms can lead to double, and even multiple, dividends and benefit the economy and the environment and even society and culture.

Economic instruments are considered to be the most efficient means to promote environmental and resource-efficient products, technologies and services to reach ambitious targets for the reduction of greenhouse gas emissions, improvement of energy efficiency and renewable energy targets in the EU. The elimination of all significant technical barriers to trade in renewable energy and environmental technologies, notably through the adoption of European standards and global trade policies, is necessary to accelerate the market introduction process (demand pull) together with technology development (technology push).

Eco-taxes can improve the reward structure throughout the economy and lead to the mutual reinforcement of environmental and fiscal policies. Other economic instruments include liability systems and performance bonds, pending satisfactory environment performance, and transferable development rights, when the landowners in sending areas are given development units, to be exercised exclusively in the receiving zone.

Voluntary approaches confer to both industry and public authorities a positive image and spur the active participation of the local society. Encouraging public reporting by enterprises on environmental performance can accelerate progress. The World Resources Institute encourages eco-responsible businesses to "walk the talk" and plan and develop their greenhouse gas inventory and manage greenhouse gas emissions (WRI 2006).

Eco-labelling schemes inform consumers of the environmental characteristics of products and processes. They encourage businesses to market products and services that are kinder to the environment by awarding them a logo with a visual identity, such as the European flower or the Nordic swan and the green keys in Amsterdam.

The EU Eco-Management and Audit Scheme (EMAS) is a management tool for organisations that accept they should assess, report on and continuously improve their environmental performance. Consumers and shareholders are asking for environmentally friendly products and services that are delivered by socially responsible organisations with visions, strategies and operations in line with sustainable development. The EMAS scheme has been available for participation by companies in the EU since 1995 and was originally restricted to companies in industrial sectors. Since 2001, the scheme has been open to all economic sectors, including public and private services. Participation in the EMAS is voluntary and extends to public and private organisations operating in the EU and the European Economic Area.

To receive EMAS registration, an organisation must undertake an environmental review considering all environmental aspects of its activities, products and services, its legal and regulatory framework and existing environmental management practices and procedures. In the light of the results of the review, the organisation has to introduce an effective environmental management system aimed at achieving the organisation's environmental policy and set priorities, objectives, means, operational procedures, training, monitoring and communication systems. An environmental audit should assess the management system in place, in particular its conformity with the policy, as well as its compliance with relevant environmental regulatory requirements. Finally, the organisation should provide a statement of its environmental performance and present the results achieved against the environmental objectives and the future steps designed to continuously improve its environmental performance.

Improving eco-efficiency (economic and ecological) is a promising path for visionary and committed cities, governments, industry and citizens willing to associate double dividends with sustainable development. The Business Council for Sustainable Development (later the World Business Council for Sustainable Development, WBCSD) adopted eco-efficiency as a business concept in 1992, in its report to the Rio summit. Eco-efficiency requires doing more and better with less impact on the environment. It is achieved by the delivery of competitively priced goods and services that satisfy quality of life, while progressively reducing resource intensity and ecological impact throughout the life cycle. Energy intensity, material minimisation, service intensity of goods, minimisation of the toxic dispersion, improvement of the life cycle and maximisation of the use of renewable resources are key criteria for eco-efficiency (Fussler 1996; WBCSD 1998).

Eco-innovation is a key route to eco-efficiency. It can be supported by various instruments, including voluntary agreements and approaches, such as extended producer responsibility, and environmental management systems which encourage changes in resource inputs and the complete remodelling of products and processes. Eco-design, design for the environment and the economy, takes environmental considerations into account in product and process engineering and marketing. Eco-companies, reducing the impact of processes and products throughout their life cycles, are expanding.

Although important improvements in eco-efficiency have been achieved over the last few years, the increasing demand for resources outweighed the eco-efficiency gains. Improving eco-efficiency throughout the economy demands actions such as the adoption of input and output indicators at a national level and the promotion of eco-efficiency economy-wide. Strategies for eco-innovation include efforts for sounder household and mobility consumption patterns, eco-education, information and communication. Adoption of best practices can improve performance and benefits.

Environmental technologies are gaining ground fast in the EU. The further development of the Emissions Trading Scheme is expected to increase their share of the market. The adoption of EU-wide standards, labelling requirements and common rules, for example on green procurement, are also crucial. The wide acceptance of early developed EU environmental standards could lead to subsequent gains in global competitiveness.

Green procurement can uphold sustainability priorities. Greening public procurement rules at EU and national level could be an important means of substantially reducing unsustainable production and consumption patterns and could be instrumental for promoting low-carbon technologies in the market. Urban examples include hydrogen buses for public transport and green buildings for municipal schools, hospitals and city administrations. Public authority spending in the EU amounts to around 16% of GDP. Alliances among public authorities for joint green procurement could impact greatly on the size of the markets and achieve crucial price reductions.

In 2006, the renewed EU Sustainable Development Strategy included the goal of bringing the average level of green public procurement in the EU up to the standard achieved by the best-performing member States by 2010. A proposal to set ambitious targets for green public procurement, as part of a broader action plan for sustainable consumption and production, has been adopted and called on governments to make sure that half of all their tendering procedures comply with a set of common green criteria by 2010. The proposed 50% target would be indicative only and member States could set their own sectoral targets to adopt even more ambitious environmental criteria.

Cities could do much to overcome barriers preventing more widespread take-up of green public procurement, including the perception that green products are more expensive and a lack of awareness of the benefits, insufficient information and training both for administrations and consumers, deficient political and managerial support and the lack of harmonised procedures and criteria across the EU.

Joint municipal green public procurement across Europe could be based on common criteria to help avert distortions of the single market and reduce the administrative burden. The minimum technical specifications could be based on consultations with industry and civil society as well as on existing European and national standards. Cities could adopt labelling standards below which public authorities would not be allowed to procure goods. The identified priority sectors for greening public procurement, on the basis of their importance and scope for environmental improvement, public expenditure, impact on the supply side, existence of relevant criteria, market availability and efficiency, include construction, food, transport, energy, office machinery and computers, clothing, paper and printing services, cleaning and health.

Ethical citizens and vigilant consumers call for cities and companies to design and prove their contribution to sustainable development. It is estimated that 80% of all product-related environmental impacts are determined during the product design phase. An EU framework Directive on eco-design establishes requirements for life-cycle approaches as early as possible into the product development process.

Awareness campaigns can overcome information deficits and increase the consumer's knowledge on the ecological impacts and the potential benefits of alternative consumption patterns. Citizen associations and the media tend to create a climate of trust, surrounding sustainable cities and businesses. Companies and municipalities without a declared commitment to sustainable development may face consumer boycotts, attacks on fixed assets, failure to attract environmentally sensitive stakeholders, stockholders and employees, expenses to remedy past mistakes, restrictions on operations and obstacles in raising finance and insurance. A proactive approach is a must, since a damaged reputation, impaired licenses, disillusioned shareholders and disappointed citizens may impose a disproportionate a posteriori cost.

A growing number of cities and companies are preparing for zero emissions. Like urban declarations, corporate codes of conduct have gained momentum. Assessment and public reporting is a necessary complementary component of the whole process. The EU Strategy on Sustainable Development invited all publicly quoted companies with at least 500 staff to publish a triple bottom-line report and present their performance against economic, social and environmental criteria. The voluntary non-binding nature of most codes is often related to the absence of any form of independent auditing, even if the codes spell out the necessity for monitoring, assessment and reporting. Businesses adopt a broad range of approaches, from non-reporting to social reporting (SustainAbility/UNEP 1998).

Corporate engagement and disclosure is highly promoted by CERES, a leading US coalition of investors, environmental and public interest organisations and advocacy groups with a lofty 20/20 vision. The coalition advocates honest accounting, higher standards of business leadership, bold solutions to accelerate green innovation and smarter policies that reward sustainability performance. The global reporting initiative, launched in 1997, had already proposed a harmonised public disclosure to deliver a steady flow of consistent, comparable and verifiable information. CERES's fourth sustainability report enabled the organisation to "walk the talk" and serve as a model for other non-profit organisations and small businesses looking to improve their own transparency, illustrating that reporting can be a beneficial process for any organisation, regardless of size or type (CERES 2008).

Eco-auditing constitutes a valuable instrument on the road towards sustainability. Eco-auditing schemes were first adopted by private sector organisations as management tools to assess, report on and improve environmental performance. Their backbone is a systematic environmental monitoring considering all aspects of an organisation's activities and services, products and processes. Environmental auditing in the public sector and especially among local authorities is a rapidly expanding instrument for challenging urban performances. The field is still in a state of flux and it is difficult to provide a well-defined paradigm out of hybrid methods.

Transparency, credibility and accountability are key components of urban auditing schemes. Cities and enterprises often adopt parallel paths when conducting their environmental auditing. It is essential to have the rigour associated with financial auditing. Diagnosis of problems should be accompanied by prognosis of trends and adequate design of policy measures. The environmental balance sheet of Sundsvall, in Sweden, including the accounts of stocks and flows of environmental resources, can provide a model, whereas in the UK, the environmental auditing in Kirklees offers a horizon of lessons. From the internal auditing of the municipality to the external auditing of the community, urban eco-auditing can help cities become aware of their consumption and design better policies.

2.2   In Search of Innovative Ecological Models

Many cities are in search of sound and meaningful ecological models linked to socio-economic progress. Since the 1970s and especially after the Rio summit, urban ecology offered them new visions. Awareness of environmental quality has increasingly been regarded as a civic value and more cities have striven to adopt environmentally friendly lifestyles. Ecological actions are being developed through innovative partnerships rooted in the local culture. Eco-cities are not simply cities with parks, grass roofs, timber frame constructions, renewable energy systems and improved resource and water cycles. They are cities that pertinently integrate environmental, social and economic policy objectives in all their policies and search for multiple dividends to improve their overall performance, based on strong citizen participation.

Stockholm was selected, among 35 cities which competed for the prestigious title of European Green Capital of 2010, as the first Green Capital of Europe, for 2010, and Hamburg was selected as European Green Capital for 2011. The candidate cities presented a wealth of strong suggestions for improved urban environments. The eight front-running cities, including also Amsterdam, Bristol, Copenhagen, Freiburg, Münster and Oslo, could be excellent role models for green capitals all over Europe and the world.

The selection of the European Green Capital was based on ten key indicators and not only considered the cities' environmental status, but also their future plans and prospects. The ten indicators were public transport, green public spaces, air quality, noise pollution, waste management, water use, wastewater management, sustainable land use, environmental management and local contributions to the fight against global climate change.

Already during the last decade of the previous millennium, the "ecological city" concept, in the framework of the eponymous OECD project, advocated the integration of ecological concerns in all urban policies. The process was described as an essential bridge between the macro-level concept of sustainable development and the micro-level concept of everyday local performance (OECD 1996).

The Liveable Communities Awards, launched in 1997 and endorsed by the UN Environment Programme, focused on best practice regarding the management of the local environment. The objective is to improve the quality of life of individual citizens through the creation of sustainable liveable communities. Peer learning and joint forces to address mutual challenges have been critical. In the Netherlands, the concept of the liveable city, together with the interrelated ones of the well-ordered city, the affordable city and the sustainable city, created a tetraptych for the vision for the sustainable city.

Leading world examples include the promotion of an eco-society by the Tokyo metropolitan government to advance towards a clean, sound and citizen-friendly metropolis. The action plan includes comprehensive actions on sustainable resource and water management, energy, transportation demand management and promotion of environmental education and awareness (UN/Tokyo Metropolitan Government 1998).

Urban ecology has been an important element for the "renaissance" of reunited Berlin, which, 20 years after the fall of the wall, pulses with energy and vibrancy. In one of the most creative quarters, Kreuzberg, during the years of euphoria, various building blocks were convincing examples of a new era for the city. In Block 103, former squatters were given the opportunity to own the space they occupied, if successfully trained in converting it into environmentally friendly and resource-saving buildings. Another complex, Block 6, was the innovation field for alternative water systems. The project offered inhabitants the possibility to monitor their hydraulic resources and take good care of their rainwater and grey water. The process emphasised communication and learning and resulted in important water savings. Eco-stations have been created in many neighbourhoods for awareness raising, training and counselling (Gelford et al. 1992).

The key driving principles were that nothing is impossible and everybody has to participate. Schwabach, a small self-sufficient German city, transformed into an ecological laboratory, offered an example for experimenting with new concepts and actions to elaborate an urban ecology planning strategy. The city, selected by the Federal Ministry for the introduction of ecological processes under ordinary conditions, undertook an analysis of its ecological assets and formulated a concrete 1993–2003 model urban development strategy, leading to Schwabach Ecological City (Schmidt-Eichstaedt 1993).

Leicester, the first UK winner of the "Environment City" award, offers a plethora of ideas. The city, assisted by the Business Sector Network, brought together ideas from the economic sectors. Environ, a non-profit-making company, helped local organisations with environmental audits and provided advice. The energy efficiency centre promoted action for improving the efficiency standards for buildings and schools, introduced an energy education package for teachers and invited students to contribute to the energy monitoring of their schools. The energy efficiency bus, equipped with solar panels, has promoted awareness on renewable energy among schools and enterprises (Energie-Cités 2001).

Ecology was a major instrument of socioeconomic change during the reunification of Germany. In Leipzig, non-governmental organisations, in cooperation with the municipality, launched environmental projects to improve the living conditions in a city in metamorphosis. The ecological restructuring of parts of Leipzig offered a good example of the regeneration of the urban fabric through ecological projects and the establishment of natural links between the city and the countryside. Certain projects achieved optimal results. Urban transport improved and attractive green corridors connect green spaces inside and outside the city (Hahn 1997).

To promote the exchange of the local authorities' experiences on sustainable development, the Association of Mayors of France, the association 4D, conceived in 1993 as a citizen's network for the monitoring of the commitments by France and the other member States of the UN, the ministry in charged of ecology and Committee 21, a network of all those involved in sustainable development, created, in 2006, a national observatory of local Agenda 21 plans and the territorial practices of sustainable development.

This observatory wishes to serve as a tool of networking and supporting instruments devised to achieve progress towards sustainable development. Valuable information, offered voluntarily by the local authorities on the basis of a common questionnaire, is communicated to a data bank with free access on the Internet. The information relates to the participation processes of the various actors, the drivers and obstacles for change, the conceptual documents and the operational tools (diagnosis, action plans, local Agenda 21 plans, booklets, dashboards).

Swedish local authorities were among the first to adopt local agendas and Agenda 21 plans by the targeted deadline in Rio (1996). The plans included comprehensive actions on resource and waste management, transport, consumption patterns and environmental education. They are essential for cities trying to create an eco-culture and conceive, introduce and manage new ecosystems, founded on active citizen participation.

French cities have introduced urban environmental plans during the last 20 years. The preamble to the French constitution refers expressly to the charter of the environment and sends a strong message. The French urban environmental plans, introduced in the 1990s, comprise, in general, an exhaustive diagnosis of the present situation, a forecast of the tendencies based on scenarios and proposals for joint action. The most frequent actions are the protection of the natural resources, and the improvement of public health, of urban safety and of living conditions.

Recreating cities as vibrating ecosystems and places of citizenship is at the heart of the debate on the sustainable city (Mega 2005). Consume better, move differently, reduce exclusion, encourage an improved model of development, revivify democracy and live better were the six objectives of the Agenda 21 of Orleans. The agenda included 234 proposals articulated around the social, the environmental and the economic pillars of sustainable development. It also rested on two important principles, the interoperability between the municipal services and the participation of the largest number of citizens.

Unlike nations or even regions, cities are sited in specific climates with distinct geographical features. Wind, solar and tidal energy and locally produced biofuels capitalise on geographical differences. Organic food development and green building approaches are also the result of regional geographies and climates. Lifestyles and architecture can be further enriched by local cultural and historical preferences and knowledge.

A new district of ecological residential buildings created in the suburbs of Mannheim is an excellent example. The urbanisation of a green area created an air-conditioning corridor for the entire city. A district has been conceived in an ecological spirit with an aquatic park and an idyllic atmosphere. Social mix has been obtained thanks to the mixture of social housing and private housing, occupied by their owners and tenants, and public and green spaces. Urban promenades completed a powerful urban experience.

Fredrikstad was the first municipality in Norway to adopt an all-party plan of environmental protection. The city has Norway's first modern incineration plant for the advanced flue gas treatment of solid waste. Energy from the plant is sold to nearby processing industries, dramatically reducing the use of oil. The preservation of the old town and the protection of the natural surroundings including Ora, one of the country's most important wetlands, are exemplary. The environment forum bringing together all major stakeholders, promotes cleaner production, a green public sector, environmental monitoring of natural assets and environmental education. The Fredrikstad environmental forum insists that environmental care pays off and wishes to contribute to a change of mentality from "use and throw" to "throw and use".

Good environmental gestures are learnt early in life. Education for sustainable development is high on the agenda of cities, governments and international organisations. The initiative "Hamburg is Learning Sustainability" makes Hamburg an inspiring role model. Hamburg, Germany's second largest city and Europe's second largest port, has 13% parks and 7% nature reserves. Environmental tours in the city, a green film festival, the green youth summit and specialised events and workshops demonstrate a particular environmental sensitiveness. In 2006, Hamburg was the first city in Germany to be awarded the title "Official City of the UN Decade Environment for Sustainable Development 2005–2014". The publication series "Moving Worlds Through Education" and the quarterly newsletter shed light on a range of important actions. The Hamburg plan of action includes a catalogue of about 120 measures, including actions for all educational areas and standards for inclusion in best practices. Intercultural learning, fair trade, global solidarity, green mobility and sustainable investment are high on the agenda of "Future Weeks 2009", organised in cooperation with the future council of Hamburg, a public forum of institutions, associations and enterprises.

Many world schools promote high environmental standards and cultivate the sensitivity of young pupils to water, energy and waste issues. The international award programme "Eco-Schools" helps schools on their sustainable journey and provides a framework to embed sustainability principles into school life. Eco-Schools is one of five environmental education programmes run internationally by the Foundation for Environmental Education, which runs also Green Key, Young Reporters for the Environment, Blue Flag and Learning About Forests. Forty-six countries participate in the Eco-Schools programme, linking more than 40,000 schools. The UK Government wants every school to be a sustainable school by 2020.

Participation in the Eco-Schools programme makes tackling sustainable issues manageable for all schools, including nurseries, primary schools, secondary schools and schools with special status. Once registered, schools follow a simple seven-step process which helps them address a variety of environmental themes, ranging from litter and waste to healthy living and biodiversity. Children are the driving force behind the project; they lead the eco-committee and help carry out an audit to assess the environmental performance of their school. Through consultation with the rest of the school and the wider community, pupils decide the environmental themes that they wish to address and the ways to do this. Measuring and monitoring is an integral part of the Eco-Schools programme.

Religions can also play an important role. UNEP in partnership with the Alliance of Religions and Conservation organised the interreligious conference "Many Heavens, One Earth", in November 2009, where many of the world's faiths presented 7-year plans for greening their activities. Many religious groups hold land and buildings and run parishes, schools and hospitals and perform functions with an important carbon and environmental footprint. The Church of England has pledged to reduce carbon emissions by at least 42% by 2020 and by 80% by 2050. Under the Muslim 7-year plan, the holy city of Medina is expected to become a model green city.

Symbolic events hold much potential for introducing innovations and influencing everyday decisions. The annual editions of "Urban Green Days", celebrated by hundreds of cities and towns in the EU, under the banner of "Green Week" during the World Day of the Environment, offer greenhouses of ideas and actions for green living, with better homes and neighbourhoods, organic food and waste, green travel and green reporting. They can provide an excellent opportunity for cities to open their doors, bring together their inhabitants and discuss their assessments and their projects for a more sustainable future. Participating cities often go beyond the state of the art. Awards link recognition with awareness raising. Annual festivals can be very instructive regarding collective awareness. London used the Urban Green Days to launch a new Web site, mapping community-based environmental events across the city.

2.3   Lighthouses of Actions for Climate

Climate change, the most prominent and urgent issue on the global sustainable development agendas, is a multifaceted and pluridomain phenomenon. During the last decade, climate politics steadily climbed up the planetary agenda, attracting significant attention. Climate change represents a planetary challenge with important intra- and intergeneration equity dimensions, already affecting developing world citizens and threatening future generations. Cities and countries that have caused the bulk of emissions are not those most likely to suffer their worst impact. Industrialised countries have a responsibility for leadership. Gobal participation, with common but differentiated responsibilities, is required for a lasting solution (OECD 2003; Garvey 2008).

Addressing climate change requires new alliances, new mindsets, new frameworks and new spaces for interaction. Bolder policy portfolios, including regulation and enforcement, economic instruments, voluntary measures, research, innovation and education, information and awareness raising, are being devised for the protection of the climate and the adaptation to climate change. Alliances bringing together stockholders, stakeholders, shareholders and citizens can play a critical role in reducing emissions (Gore 2007).

Emission patterns are influenced by long-lived investments in energy production and consumption, transport infrastructure, housing and industrial installations, which impact lifestyles on a long time horizon. Immediate change is difficult and technological, socioeconomic and institutional innovation is crucial for inefficient patterns to become smarter and healthier. New concepts, values and technological breakthroughs, linked to scientific developments, policy and market initiatives and public expectations, are already challenging the inertia of old infrastructures and outmoded patterns (Stern 2009).

The UNFCCC issued during the 1992 Rio conference and enjoying near universal membership, recognised the challenge of climate change. Furthermore, it set an ultimate objective of stabilising greenhouse gas concentrations in the atmosphere at a level that would prevent dangerous anthropogenic interference with the climate system. At the Third Conference of the Parties in Kyoto in 1997, the Kyoto Protocol to the UNFCCC marked an important milestone. It is a pact among industrialised countries, committed to significant reduction of emissions and unambiguous targets for greenhouse gas emissions, which introduced three flexible instruments, emission trading, joint implementation and the clean development mechanism. The protocol entered into force in 2005 and is considered the first step in a long struggle to tame the global climate (OECD 2000c).

In 2007, the IPPC presented the Fourth Assessment Report evaluating the scientific knowledge of the natural and human drivers of climate change, observed changes in climate, the ability of science to attribute changes to different causes, and projections for future climate change. The report, approved by all governments of the world, stated that the warming of the climate system is unequivocal and most (over 90%) of the observed increase in globally averaged temperatures since the mid-twentieth century is very likely due to the observed growth of anthropogenic greenhouse gas concentrations (IPPC 2007).

The last IPPC report confirmed the unnatural rise of greenhouse gases, including CO_2, methane (CH_4), nitrous oxide (N_2O) and the three main fluorinated gases, hydrofluorocarbons, perfluorocarbons and sulphur hexafluoride (SF_6). The amount of CO_2 in the atmosphere in 2005 (379 ppm) exceeded by far the natural range of the last 650,000 years (180–300 ppm). The primary source of the increase is fossil fuel use, but also land-use changes. The amount of CH_4 in the atmosphere in 2005 (1,774 ppb) exceeded by far the natural range of the last 650,000 years (320–790 ppb). The primary source of the increase in CH_4 is very likely to be a combination of human agricultural activities and fossil fuel use. N_2O concentrations have risen from a preindustrial value of 270 ppb to a 2005 value of 319 ppb and more than a third of this rise is due to human activities, primarily agriculture.

Evidence of climate change has grown during recent decades, both on land and in the oceans. During the last century, the average global temperature increased by 0.74°C. Average Northern Hemisphere temperatures during the second half of the twentieth century were very likely higher than during any other 50-year period in the last 500 years and likely the highest in at least the past 1,300 years. The average Arctic temperatures increased at almost twice the global average rate in the past 100 years.

Sea level rise is also most often associated with climate change, as ocean warming and melting of snow and ice cause the sea level to rise. During the last century, the average global sea rise was 17 cm. During 1993–2003, there was an average rise of 3.1 mm/year (against 1.8 mm/year during the years 1961–2003) and if present trends continue, projections suggest an increase of 9–88 cm by 2100. Extreme weather events, including storms, floods, heat waves and droughts, have been more frequent. There has been an increase in hurricane intensity in the North Atlantic since the 1970s, in correlation with increases in sea surface temperature (IPCC 2007).

A change in the climate could potentially alter the conditions that govern human life and lead to major costs. The global community should aim at reducing emissions of CO_2 and other greenhouse gases to acceptable levels as rapidly as possible. On the basis of scientific evidence, the EU has reached political consensus that +2°C above preindustrial levels is the maximum safe level for humanity, taking into account the precautionary principle. The risks of the worst impacts could be prevented if greenhouse gas emissions in the atmosphere are stabilised between 450- and 550-ppm CO_2 equivalent. The cost of action can be limited to 1% of global GDP each year and it is estimated to be far less than the cost of inaction. Furthermore, the world does not need to choose between averting climate change and promoting prosperity (Stern 2006, 2009).

Power supply is the single most important contributor to greenhouse gas emissions and it is expected to remain so in the near future. Deforestation contributes to releasing CO_2 into the atmosphere and conversely geological sequestration helps to diminish the amount of CO_2 in the atmosphere. The loss of natural forests contributes more to greenhouse gas emissions each year than the transport sector (Stern 2006).

The EU (then EU15) committed itself to reducing its greenhouse gas emissions to 8% below the 1990 levels by 2008–2012 and it is leading world efforts for more substantial reductions by 2020 and 2050. The last annual submission of the greenhouse gas inventory of the EU to the UNFCCC, presenting greenhouse gas emissions by individual member States between 1990 and 2008, suggests that emissions in 2008 had fallen for the fourth consecutive year. According to the European

Environment Agency, emissions by the EU15 fell by 1.3% in 2008 compared with official figures for 2007 and by approximately 6.2% below the Kyoto base year (1990 in most cases). This is encouraging as it seems to indicate that the EU is on target to meet its Kyoto commitments for the period 2008–2012. In fact, the European Environment Agency estimates that the emissions from the EU27 fell by 1.5% in 2008 to a level that is 13.6% below that of the base year (EEA 2009a).

The cost-effectiveness of policies to fulfil the Kyoto commitments is an important element of the EU strategy. The European Climate Change Programme (ECCP), set up by the European Commission in 2000, tried to identify the most cost-effective additional measures for enabling the EU to meet its target, complementing the efforts of member States (EC 2003b). In 2005, the second ECCP reviewed the first phase, focusing on transport, energy supply, energy demand, non-CO_2 gases, agriculture, and proposed actions on aviation, CO_2 and cars, carbon capture and storage, adaptation and the EU Emission Trading Scheme. The policies and measures under the ECCP have a total emission reduction potential twice as high as required for the EU15 to achieve its target.

Emission trading is a flexible mechanism that is attractive for governments wishing to reach their targets at significantly lower costs. The instrument confers more flexibility and cost-efficiency than direct regulation and may offer a higher degree of effectiveness than other instruments, such as taxation and voluntary agreements. Denmark and the UK introduced national emission trading schemes in 2001 and 2002. In 2005, the European Commission introduced the groundbreaking EU-wide Emission Trading Scheme, after the acceptance of the first national allocation plans, including the number of CO_2 emission allowances (rights to emit) that member States allocated to selected energy-intensive industrial plants according to transparent criteria.

The 3-year mandatory initial phase (2005–2007) of the Emission Trading Scheme has provided valuable lessons for the improvement of the system and the national allocation plans for the second trading period, 2008–2012. The second phase (2008–2012) foresees an annual EU-wide cap of 2.08 billion allowances, half of the EU total emissions. The EU27 national allocation plans could bring about a 6.5% reduction below the 2005 levels. The aviation sector will be included from 2011 onwards, and the extension to other greenhouse gases, other than CO_2, is also being discussed. Ex ante benchmarking for the free allocation and auctioning may be important to avoid undue distributional effects. The links to the other flexible instruments, Clean Development Mechanism and the Joint Implementation, and the gaining of credits through the support of projects in developing or in other industrialised countries and economies in transition can bring further opportunities (OECD 2007b).

Global warming is partially the end product of billions of individual decisions made by the world's residents within their immediate environments. The consequences of climate change differ largely throughout the world. Therefore, both prevention and remediation of effects should be worked out and implemented locally. Cities have a cardinal role to play. Many climate-conscious cities have undertaken noteworthy initiatives for averting climate change.

Elected mayors were among the first to be mobilised in the fight against global warming. The World Mayors' Council on Climate Change, an alliance of committed local government leaders, initiated by the then mayor of Kyoto in 2005, following the entry into force of the Kyoto Protocol, advocates an enhanced recognition and involvement of mayors in multilateral efforts addressing climate change and global sustainability.

The Climate Alliance of European Cities consists of an association of around 1,400 cities, municipalities and districts and more than 50 provinces, non-governmental organisations and other organisations. The member cities and municipalities aim at decreasing greenhouse emissions. To achieve this goal, they devise and implement local climate strategies. Furthermore, measures are being implemented to raise public awareness on the protection of the rainforest and municipal procurement and abstaining from tropical timber derived from destructive logging. The partner of the alliance, the Coordinating Body for the Indigenous People's Organisation of the Amazon Basin, representing the interests of the indigenous rainforest population, brings an important element in the South–North ethical debate.

Local governments in Europe have adopted ambitious targets and comprehensive action plans in support of the EU's climate goals which often go beyond the national targets. The Local Government Climate Roadmap aims to support efforts to achieve a comprehensive post-2012 global climate agreement (EC 2009b).

The Cities for Climate Protection (CCP) campaign, one of the key ICLEI's programmes, was introduced in 1993. More than 130 local authorities in Europe have joined the CCP-Europe campaign and more than 440 worldwide, accounting for approximately 5% of global emissions. The mission of the CCP campaign is to build and serve a worldwide movement of local governments which adopt policies and implement measures that achieve measurable results for local greenhouse gas emissions, improve air quality, and enhance urban liveability and sustainability. The campaign offers a framework for local authorities to commit and develop climate protection policies and actions.

An important first step to reduce greenhouse gas emissions is for each citizen, household and economic actor to become aware of their carbon footprints. The Davos Climate Alliance is a valuable tool to become more conscious and for compensation. It was introduced during the 2009 forum and strives to make the World Economic Forum Annual Meeting carbon-neutral. Through the Davos Climate Alliance, participants can compensate for their personal greenhouse gas emissions by contributing to an offset programme investing in a project promoting climate-friendly technology. This initiative highlights that unmitigated climate change poses a threat to the global economy and community. It could inspire cities wishing to offset the carbon footprint of all their international gatherings.

Individual cities prepare their climate plans. Amsterdam has confirmed its will to become climate-neutral by 2015 and Malmö intends to be a climate-neutral city by 2020 with respect to municipal sector activities. Its close links to Copenhagen, through common policies, economic ties, cultural relations and a physical connection, via the Öresund Bridge, provide Malmö with a strategic opportunity to heighten joint efforts to confront climate change. As Malmö becomes a renowned resource city in terms of its attention to urban environmental planning and policy, the climate campaign is trying to build greater awareness among Malmö residents and increase visibility concerning planned activities and activities already under way (Malmö 2008).

Amsterdam set the goal of reducing the CO_2 emissions within the city limits by 40% in 2025, compared with the level in 1990. Despite the efforts of the city in the area of energy savings and the development of renewable sources, CO_2 emissions had increased significantly since 1990 and the city decided to move from a slow increase to a drastic reduction. In 2007, the Municipality of Amsterdam took the initiative to establish a framework of cooperation for climate, the New Amsterdam Climate, made up of citizens, businesses and institutions wishing to take an active leading role. New Amsterdam Climate is a platform where parties can find partners for cooperation, inspire and motivate each other, share knowledge, detect and remove obstacles, promote actions and results and present annual CO_2 reports (Amsterdam Climate Office 2008).

Every year, the City of Amsterdam monitors the situation by determining how much CO_2 the city has emitted. It also collects as much information as possible about the CO_2 emissions that have been avoided owing to the various measures. To this end, a model is being constructed and supplied with information about energy use in the built environment, transport, businesses and the city administration.

Clear vision, concrete targets and bold leadership have been the key elements of the London Climate Change Action Plan. The creation of the Greater London Authority provided the strategic capacity bringing together economic development and transport and environmental planning for eight million people under the leadership of an elected mayor.

The London Climate Change Agency, established in 2005, is working towards preparing firm legislation to reduce the impact of London on the climate and the environment and deliver ground-breaking energy-efficiency and renewable-energy projects throughout the UK capital. The Climate Change Action Plan, issued in 2007, focuses on the next 10 years in the context of achieving the 2025 target of 60% emission reduction

Even after the withdrawal of the USA from the Kyoto Protocol, in 2001, American cities declared themselves in favour of the engagement against global warming. The mayor of Seattle mobilised mayors from across the nation to join the non-partisan Mayors Climate Protection Agreement. By autumn 2008, more than 880 mayors representing over 80 million Americans had signed the act, urging the US and state governments to meet or beat the carbon reduction goals set by the Kyoto Protocol, while vowing to take local actions to reduce global climate change.

The Boston Night to Combat Climate Change was an exceptional event drawing attention to the global climate change crisis in November 2009, just before the critical negotiations. The event featured many notable speakers, networking opportunities, a silent auction, dancing and entertainment. The organisers promised a carbon footprint of the event that would be measured and offset, and participants received an event package containing Earth-friendly products and information on ways to reduce their carbon footprint. The event focused on raising awareness about climate change, while also celebrating the efforts of innovators working to address the environmental and humanitarian impacts of global climate change.

The Clinton Climate Initiative (CCI), launched by the Clinton Foundation in 2006 to advance progress against climate change, adopted a holistic approach, addressing the major sources of greenhouse gas emissions and the people, policies, and practices impacting them. Working with governments and businesses, CCI focused on three strategic policy areas, increasing energy efficiency in cities, catalysing the large-scale supply of clean energy and working to bring deforestation to an end. Through targeted projects, including building retrofits, outdoor lighting and waste management, CCI's Cities programme helps municipal governments improve energy efficiency and measure the reductions of emissions.

The CCI collaborates with the 40 larger cities of the world to reduce greenhouse gas emissions, while starting with an ambitious programme of improvements of buildings, supported by major banks. The C40 coalition of the world's largest cities is highly committed to tackling climate change. From Bogotá to Chicago and Melbourne, under the leadership of Toronto, the coalition strives for emission reductions and better energy efficiency. It invited national governments to engage more closely with their city leaders, best placed to deliver greenhouse gas emission reductions, empower city leaders so that they have the necessary authority to take action and decrease greenhouse gas emissions and resource cities so that they are well equipped with the relevant tools, services and finance to help deliver national targets.

State and regional plans are also progressing. Massachusetts committed to emission reduction targets of 25, 40 and 80% by 2010, 2020 and 2050, respectively, in relation to 1990. The US Conference of Mayors and the American Institute of Architects, recognising that the share of buildings in relation to national energy consumption and CO_2 emissions is 48%, adopted the challenge to reduce to zero the use of energy coming from fossil fuels by 2030 (MIT 2008).

At the level of enterprises, the Combat Climate Change (3C) initiative brings together more than 50 innovative firms wishing to make a difference towards a low-carbon economy. CERES launched the Investor Network on Climate Risk, a group of more than 70 leading institutional investors with collective assets of more than $7 trillion. It also published cutting-edge research reports to help investors better understand the implications of global warming and advocates the necessity of global action for climate justice and sustainable prosperity.

As the process leading up to the climate change conference in Copenhagen, host of COP 15 in December 2009, gathered speed, local politicians and officials from almost 160 countries and representatives from national governments joined the Local Government Climate Change Leadership Summit, in June 2009. This was a milestone in the Local Government Climate Roadmap, advocating a national–local dialogue and a partnership to reduce greenhouse gas emissions. The key messages of the Local Government Climate Roadmap are that the parties recognise the role of cities and local authorities, empower them and engage with them to provide enabling structures and effective framework conditions for cooperation on climate protection.

Climate change is not just an environmental issue for cities. The entire social fabric, the economy and the governance of cities are at stake and collaboration with other levels of governments is imperative. Local, regional and national authorities should be supported in their efforts to combat climate change. The national level is needed to support the action plans for climate change with economic and political tools. Transfer of knowledge between local governments is essential and financial support should facilitate this, especially in the developing countries hardest hit by all types of crisis.

Cities committed to protecting the climate have joined the City Climate Catalogue to influence international climate negotiations. This interactive tool, managed by the City of Copenhagen together with ICLEI, was launched in February 2009 and intends to provide a substantial argument for national governments in the international climate negotiations. All cities are invited to contribute their greenhouse gas reduction targets and engage in actions to achieve and exceed them.

Commitments, when followed by action, monitoring and reporting, can help build a strong case for local climate action and encourage cities to stand behind their commitments. The aim is to capture and disseminate useful information on as many local targets as possible. In a second phase, the objective is to collect and share information on interesting local developments and effective climate protection implementation actions that lead to substantial greenhouse gas reductions in the catalogue. The focus is on energy savings, energy efficiency and the use of renewable energy sources.

A declaration on climate change from some 130 major cities across 34 European countries testifies to the commitment of cities to taking local action on climate change. The declaration provides guidelines for implementing local policies to reduce greenhouse gas emissions by involving all local actors and engaging in better urban planning, transport, renewable energy and diversified energy production. Examples of the proposed actions include the adoption of ambitious sustainable public procurement policies, innovative partnerships in the fields of research and education, promotion of renewable energies, urban sprawl control and the development of compact cities, construction and rehabilitation of energy-efficient buildings and the creation of new eco-quarters.

There is, however, increasing evidence that climate change cannot be prevented and adaptation measures are also crucial. Cities have to engage in adaptive actions to keep the impacts of climate change within manageable boundaries, apart from the reductions in greenhouse gas emissions to stabilise temperature rise. With a changing climate, extreme events are likely to occur more frequently. With their high population concentration and physical structure, cities and towns are extremely vulnerable to these threats.

Many coastal cities face a serious risk of flooding as sea levels are expected to rise. Droughts and heat waves represent other major threats. The likely impact of extreme events related to climate change depends on location and morphology, but also the physical characteristics and city design and planning, as has been witnessed with the urban heat island effect, which is caused by differences in urban density and vegetation cover.

Many cities may suffer considerably, in terms of both population and environment, with significant socioeconomic implications. In the developing world cities, these impacts may aggravate poverty conditions and social inequalities. Climate change will also exacerbate other existing urban problems such as low air quality and poor water supply. Investing now in mitigation and adaptation will help to avoid huge costs later. Some cities already see this as an opportunity for better urban planning and policy to improve the quality of life and create new employment opportunities.

Initiatives at the city level also have to link with actions at the regional, the national and the European level. For instance, cities vulnerable to drought or excessive rainfall need to act in tandem with their surrounding regions to increase water storage capacity. Some countries, such as the Netherlands, can draw on extensive experience of protecting their coastlines and cities and provide know-how and best practice. Urbanised artificial islands on reclaimed land become laboratories for flood-resistant concrete blocks moving vertically up and down. The Dutch Delta Committee Recommendations include a comprehensive action programme to secure a climate-proof Netherlands and safeguard coasts, urban areas and the hinterland up to 2050.

Cities initiating early measures are expected to see the best returns on their adaptation investments. London's Climate Change Adaptation Strategy proposes action far beyond the Thames barriers and tidal mechanical defences. The measures include management of the flood risk from tributaries to the Thames and heavy rainfall. The existing Thames barrier and tidal defences are expected to protect London for decades to come and it is unlikely that a new barrier would be required in the foreseeable future. Copenhagen's proposed Climate Plan includes an adaptation plan that should also create environmental initiatives and continue to improve the city's recreational opportunities. The city opted for the creation of pocket parks, i.e. small green spaces integrated in the urban fabric which help cool the city on hot days and absorb rain on wet days.

The EU is already working on an online knowledge management tool (EU Clearing House Mechanism) to share and manage information on climate change impact, vulnerability and best practices on adaptation (EC 2009b), which is also expected to provide valuable guidance for cities.

2.4   The Fundamental Urban Resources

The fourth assessment of Europe's environment, by the European Environment Agency, suggests that raising greenhouse emissions reflect our inability to live sustainably. They highlight that the area of built-up land in Europe is growing much faster than the population and that urban sprawl may have disastrous effects on the environment (EEA 2007a).

Land, air and water are the primary resources for any city. Urbanisation increases the demand for land and the pressure on wild land and results in air and water pollution. Forests are being changed to agricultural land or urban areas. Soil is a vital and largely non-renewable resource. Soil erosion and the decline in soil quality is a major and often irreparable problem across the EU.

A quarter of the EU's territory has been directly affected by urban land use. During the last decade of the century, an area five times that of Greater London was consigned to urban sprawl. This primarily occurred on former agricultural land, resulting in the loss of important ecosystem services, such as food production, flood protection and biological diversity (EEA 2006).

Urban sprawl has started to be contained in most European cities but it is still a major issue in cities of the former eastern and central Europe. In Prague, the Velvet Revolution sparked a 20-year building boom. Every year, thousands of hectares of orchards and farmland are paved over to make way for new housing developments and shopping centres (EEA 2009c).

Soil, defined as the top layer of Earth's crust, is formed by mineral particles, organic matter, water, air and organisms. Soil erosion and contamination are the other main dangers. It is a non-renewable resource and an extremely complex, variable and living organism at the interface between the earth, the air and the water. Soil performs many vital functions and serves many human activities, most of which are concentrated in cities.

Sealing of soil surfaces owing to increasing urbanisation and new infrastructures is the main cause of soil degradation in most industrialised countries. Soil loss by erosion is the main cause of soil degradation in the Mediterranean region. In some areas, soil erosion is irreversible, whereas in others nearly complete removal of soil has been observed. The retreat of the shoreline affects many coastal regions.

Coastal cities and zones, which attract a high percentage of the planet's inhabitants and tourists, are under increased stress. Biological diversity is ever more threatened. Devising efficient, effective and equitable land development policies, incorporating environmental imperatives, is a major challenge for decision-makers and planners.

Freshwater, the blue gold, is a vital and scarce natural resource. Unlike oil, it cannot be replaced. In many world cities, citizens have to buy their water. In Europe, many cities experience water shortages or are supplied with groundwater, the quality of which is seriously threatened. Maintenance of distribution networks remains a major concern. Leakage and risk detection are increasingly parts of integrated management and early-warning systems. Renovation of networks and surveillance systems to limit leakages, sometimes reaching 30–40%, is under way or planned in many cities. In Tokyo, the setting up of a system for the early identification of leaks has reduced water losses to 9%.

Water is gradually coming to feature on the global political agenda: 1.1 billion world citizens do not have access to clean water and 2.6 billion do not have access to sanitation systems. These figures may be much higher in 2025. Between 1970 and 2000, fresh water available per person living on Earth decreased from an average of 12,900 m^3 to less than 7,000 m^3. Drought and floods are both among the extreme events accompanying climate change. Rather than water-related conflicts, one can expect policies aiming at safeguarding water quality and at financing access to drinking water.

Clean and reliable water provision is also a growing concern for EU cities, particularly in Europe's Mediterranean basin, where city dwellers must compete with farmers for access to freshwater. A recommended practice is to introduce strict metering measures and water pricing, as is already done in some countries.

Ensuring adequate supplies of safe water and sanitation is a concern for both developed and developing countries. The former have connected 100% of their populations to safe water supplies and the majority to wastewater treatment, but have to replace ageing water infrastructure and comply with ever more stringent water regulations if they are to prevent excessive water leakages. For developing countries, the problem is the basic access to clean water and sanitation. One billion world citizens still lack access to clean water and 2.6 billion have no sanitation services. About 80% of all diseases in developing countries are water-related, leading to an estimated 1.7 million deaths each year. Halving by 2015 the number of people worldwide without access to safe water and sanitation services is one of the Millennium Development Goals (2000) and commitments at the World Summit on Sustainable Development (2002). Progress is slow and best practices are especially highlighted.

In the coming decades, supplies of fresh, safe water will be subject to significant pressures, and competition for scarce resources may exert a destabilising influence and result in regional conflict. Demand is expected to grow by 50% over the next 30 years. Climate change, in the form of sea level rise and extreme events such as winter flooding and summer droughts, may further increase uncertainty and the vulnerability of water resources. The population living in water-stressed areas is set to double over the period 1995–2025, and, by 2030, two thirds of the world's inhabitants may experience moderate to high water stress (OECD 2007b).

Sustainable water management entails sufficient quantities of clean water being available to support both human needs and essential ecosystem functions. Good governance requires careful consideration of the public institutions to manage water supply and sanitation systems. Policies to encourage sustainable use by agriculture, by far the largest worldwide user of water, but also industry, are particularly important. About 400 L per person per day is used in industry (two thirds for generating energy) to process food and the limitation of its consumption is crucial. Singapore has been pinpointed as a world model for successful water management (OECD 2007b).

A sustainable financing system should rely primarily on water charges, with provisions for affordable access by the poor. Water pricing is a key element of water management. Full cost recovery water charges can help generate the necessary funds for infrastructure development, renewal and maintenance, and provide incentives for efficient water use. Many countries have been moving towards water pricing that reflects the full marginal costs of water services, combined with measures that better target support to low-income users. Partly as a result of these pricing systems, per capita water use has fallen in OECD countries since 1980, and almost half of the OECD countries have reduced their total water use. Water charges applied to industrial water and wastewater treatment have also been approaching full cost recovery levels.

The management of the water environment holds many challenges for cities. A great part of Stockholm (10%) is covered by water and the many lakes and water sheds have a high recreational value. To have all waters of Stockholm meet the requirements of the EU Water Directive by 2015, the City Council adopted a water protection plan, in 2006, setting standards for cleaner water and outlining methods to achieve this. The municipality has also done much to improve wastewater treatment and reduce the impact from storm water. The wastewater is treated by advanced technology before being discharged into the inner part of the Stockholm archipelago, a sensitive part of the Baltic Sea. Stockholm Water Company has improved the water quality by cutting down by more than half the discharge of phosphorous and nitrogen since 1995.

Celebration of water reached a summit in Saragossa, during the international exhibition in 2008, a transformational event for a city at the crossroads of many cultures. Public walks and spaces were created on the banks of Erbo, the venue of the exhibition. Emblematical buildings such as the tower of water, the bridge pavilion, a pedestrian viaduct and the largest river aquarium in the world metamorphosed the urban landscape. Magic shows, puppets and the circus of the four continents attracted many young visitors. Climate change was the principal topic of the "Iceberg: Poetic Symphony". And while the circus of the sun travelled the city everyday at noon, children were taught the importance of each drop of water. The river overflowed with enthusiasm to welcome visitors to the event, a source of amazement.

Air is a part of Earth's atmosphere and one of the most important natural resources, freely shared by all. Air pollution critically affects human health and also impacts ecosystems, buildings and monuments. There is a large array of factors that cause air pollution, including the kind and the location of the pollution source, the type and concentration of emissions, meteorological conditions and the geomorphology of the region. Impacts of air pollution on human health also depend on numerous parameters, the most significant of these being the level and the length of the exposure, age and health condition.

Urban air pollution was first noticed in the middle of the nineteenth century, when British cities, front-runners of the Industrial Revolution, experienced coal smoke in winter causing a mixture of fog and smoke, known as smog. At the beginning of the third millennium, industry embraced production techniques that do not harm air quality in developed cities. The emphasis has shifted to air pollution caused by motor vehicle emissions. Waste disposal and deforestation activities can also contribute to air quality. Serious air pollution episodes in cities have been manifested in "black smoke" phenomena, the most spectacular probably being those in Cairo in 1998 and 1999, preventing there being clear daylight throughout the city. These events have been instrumental in stimulating public awareness of and citizen participation for better air quality.

Across Europe, inhabitants of urban and suburban areas are exposed to levels of air pollution that exceed the air quality standards set by the EU and the World Health Organisation. The major contributors to air pollution include sulphur dioxide (SO_2), the gas responsible for acid rain, nitrogen oxides (NO_x), carbon monoxide (CO), suspended particulate matter and certain metals. The primary source of air pollution is the combustion of fossil fuel in final energy generation, industrial processes and transport. During recent years there has been significant success in reducing certain pollutants through source control and abatement strategies and fiscal measures. Lead concentrations dropped sharply in Europe.

According to the European Environment Agency, air quality in European cities has significantly improved through the application of ever stricter European emission standards, but more needs to be done to reduce emissions of NO_x and fine particles. During the period 1997–2006, 18–42% of the urban population was potentially exposed to ambient air nitrogen dioxide concentrations above the EU limit. Eighteen to 50% of the urban population was potentially exposed to ambient air concentrations of particulate matter smaller than 10 µm higher than the EU limit set for the protection of human health, i.e. 50 µg/m³ on more than the permitted 35 days per year. Particulate matter has serious health implications (EEA 2009c).

The percentage of the urban population exposed to SO_2 concentrations above the short-term limit decreased to less than 1% and the EU limit is thus close to being met. Urban populations have also been greatly (14–61% of the total urban population) exposed to ground ozone concentrations above the target value. Ozone causes respiratory diseases and is linked to premature death. It particularly affects asthmatics, children and the elderly. In 2003, a year with extremely high ozone concentrations owing to specific meteorological conditions, the to high concentrations increased to about 60% (EEA 2007b, 2009c).

Air pollution puts a strain on sustainable urban development. Apart from its the proportion of people exposed severe local effects, urban air pollution also has profound regional and global impacts. Urban emissions are major contributors to the problems of ozone layer depletion and ground-level ozone, global warming and climate change. The effects of exposure to a cocktail of pollutants can be multiple and amplified. The heavy metals produced during energy generation by burning of fossil fuels can cause various cancers, digestive and nervous problems. Urban air pollution also causes respiratory disease and property damage. It is therefore of the highest importance that air quality in cities is monitored and improved. City air quality profiles and modelling are essential tools for air quality management (UNEP/HABITAT 2005).

Fighting acid rain through the targeted concerted reduction of SO_2 emissions can best be exemplified by the Indian city of Agra, given the corrosive effects that ambient SO_2 has on the Taj Mahal, one of the most important cultural tourism resources of the country. Many industrial sectors had to collaborate and reduce their air emissions significantly and the overall efforts led to cautious claims of success. The effects could be irreversible if due attention is not paid in the future (UNEP/ HABITAT 2005).

Noise is another problem in cities and, according to the European Environment Agency, an underestimated one. Persistent high levels of noise are associated with learning and concentration difficulties, loss of memory and irreversible damage to health. European cities have become increasingly "noisy" and citizens are increasingly affected by noise from traffic, industrial and leisure activities. In the Netherlands, the Randstad, one of the most urbanised areas in Europe, is highlighted as an example of high noise pollution, despite noise abatement measures. Road traffic is the dominant source of exposure to noise in European cities. Furthermore, noise problems are often worse in areas of high-density housing and deprived neighbourhoods next to railway lines and air traffic corridors (EEA 2009c).

The EU Thematic Strategy on the Urban Environment (EC 2006a) reported that exposure to continuous road traffic noise affected 160 million people in the EU15 (40% of the population) at a level above 55 dB, considered a serious annoyance. Furthermore, 80 million people (20% of the population) were exposed to continuous road traffic noise above 65 dB, associated with cardiovascular effects. A more recent Eurocities study indicates that 57% of the inhabitants of some European cities are living in areas with noise levels above 55 dB, and approximately 9% experience noise levels of above 65 dB. Extrapolations of these percentages all over Europe suggest that more than 210 million people in Europe are exposed to levels above 55 dB and 38 million to levels above 65 dB (EEA 2009c).

Sustainable waste management is inextricably linked to sound resource management. Paradigm shifts insist on waste prevention, minimisation and optimisation as a precious resource. Some figures highlight an alarming reality. During recent years, the amount of non-hazardous waste such as paper, plastic and metal shipped out of the EU, mostly to Asia and particularly to China, has increased dramatically. The amount of waste paper exported to Asia increased by a factor of 10 during the years 1995–2005, the amount of waste plastic by a factor of 11 and the amount of waste metal by a factor of 5 (EEA 2009b).

In the EU, the amount of municipal solid waste generated is still increasing with GDP. In 2006, the EU27 produced 255 million tonnes of municipal waste, an increase of 13% compared with 1995. This represented an average of 517 kg of municipal waste per capita, an increase of 9% over 1995. The share of municipal waste sent to landfill decreased from 62% in 1995 to 41% in 2006. Although some countries, such as Germany, the Netherlands, Sweden, Denmark and Belgium, have almost abolished landfilling of municipal waste, others, such as the Czech Republic, Poland and Lithuania, send more than 90% of municipal waste to landfills.

Waste prevention insists on action before the waste is generated, even if investments still concentrate on the recycling end. Integrated product policies shed light on the life cycle of products from the extraction of natural resources, through their design, manufacture, assembly, marketing, distribution, sale and use to their eventual disposal as waste. Eco-efficiency and eco-labelling are crucial for the prevention of waste generation. Packaging waste has to be given particular attention (ACR+ 2009).

Approximately two thirds of residential and industrial waste comprises organic materials, which break down over time. The waste stream, termed "biodegradable municipal waste", when landfilled, degrades and generates leachate and landfill gas. Recycling or biotreatment is needed to avoid these problems and to prevent dependence on landfill as a disposal option. Eventually, landfills will only be used for stabilised material. Industrial processes including food and drink preparation, agriculture, forestry and pharmaceutical manufacture are examples of processes that may produce large volumes of putrescible waste streams. Biotreatment is required to ensure that environmental protection can be assured.

A positive sign comes from the increasing amount of recycling of municipal waste, which doubled between 1995 and 2006 and reached 101 million tonnes. Germany and the Netherlands have the highest share of recycled waste, 68 and 64%, respectively. Denmark has the highest share of municipal waste (55%) that is incinerated. Energy recovery through incineration is slowly increasing and in 2005 accounted for about 9.8 million tonnes oil equivalent of energy, i.e. 1% of primary energy production, compared with 0.6% in 1995 (EC 2008e).

Waste is also an excellent source of energy. Efforts to make value out of the waste produced provide interesting lessons. Biowaste management and anaerobic digestion of organic waste to create composting and biogas are high on the research and policy agendas. The WasteEng Conference Series is a great platform for the scientists and industries in the environmental and energy sectors to share ideas and experiences and present the progress of scientific research. The next conference, in Beijing in 2010, is expected to present the state of the art on the valorisation of waste (municipal, industrial, agricultural, electronic, plastics, construction and demolition, etc.) to energy (biofuels, hydrogen, etc.) and useful materials (raw, secondary and recycled materials) and also place emphasis on the valorisation of wastewater.

Tourist cities, such Calvia, in Spain, and Rimini, in Italy, introduced partnerships with the tourism industry for the minimisation and optimisation of the waste due to tourist activities and the huge increase of the population during the summer. In Rimini, inhabitants and hotel units take an active part in the recycling of their waste. An eco-station was created and its management was entrusted to a centre for the rehabilitation of former drug addicts. Citizens were invited to contribute to the process with materials to be recycled and they were rewarded with plants.

In Curitiba, Brazil, waste is central in the "green exchange" programme, beneficial both for social integration and for the environment. Low-income families who live in shanty towns not served by municipal trucks collecting garbage bring their litter to neighbourhood centres, where they exchange it for bus tickets and food. This also results in less waste being dumped in sensitive areas such as rivers and a better life for the poor. A specific programme for children foresees the exchange of recyclable garbage for school supplies, chocolate, toys and tickets for shows. Under the "Garbage That's Not Garbage" programme, 70% of the city's trash is recycled by its residents. Apart from the environmental benefits, the money raised from selling recycled materials goes into social programmes and the city employs the homeless and recovering alcoholics in its garbage separation plant (UNDP/UNCHS 1993).

Throughout Europe, waste is increasingly regarded as a valuable resource and awards crown innovatory steps. The Royal Innovation Award, introduced by the municipality of Barcelona, crowned in 2002 the quarter of Kronsberg, in Hanover, which was created for Expo 2000, and which was conceived as an optimisation of the management of natural resources. Children were invited to waste-free breakfasts.

2.5 Eco-Buildings, Eco-District and Eco-Cities

From buildings to the district and to the city and from the city to the agglomeration and the region, all management and governance levels provide immense possibilities for ecological improvements. Eco-businesses and eco-parks, of high environmental quality and with integrated management, make it possible to reduce the overall environmental impact. Companies could, for example, organise themselves in a manner so that the waste of one becomes resources of the other. However, the addition of eco-buildings, eco-enterprises and eco-parks does not lead to eco-polises. Apart from green areas, walls and roofs, designed to provide cooling and ventilation as well as water storage and infiltration, green and sustainable cities require profound cultural changes and a citizen consensus to give a collective meaning to all technical achievements.

Bioclimatic architecture has already given excellent results of a passive eco-responsible habitat. Certified prototypes of ecological houses transform themselves into vital urban cells everywhere in Europe. The Passiv Haus in Germany or the EcoHome in England can help save precious resources. Furthermore, they are modular and flexible, luminous and healthy. The good integration into the landscape, the orientation and adaptation to the climatic parameters, the insulation and shading, the building materials and the sustainable management of water and resources, including waste, are critical for eco-performance.

More active green buildings are gaining ground all over the world. As action against climate change becomes increasingly urgent, growing numbers of citizens are looking to dramatically reduce the carbon footprint of their homes by using more active techniques. Design information about the latest sustainable materials, technologies and techniques, the best possible use of renewable energy and the best practices in action benefit from the state of the art (Enerpresse 2006; Roaf 2007).

The first emission-free hotels and factories have already been inaugurated, and the WATT dancing club in Rotterdam is promoted as the first sustainable club in the world. Its functioning is based on enhancing the energy produced from the steps of dancers. It focuses on sustainability as a creative and attractive business concept, summarised in three keywords: "people", "planet" and "profit". WATT wishes to prove that clubbing, sustainability and commerce can definitely merge into a successful enterprise. WATT's function rooms are a great profiling opportunity for companies aiming to contribute to a sustainable society. The club also features excellent public transport connections.

New York City bears witness to a growing commitment to incorporate environmental factors into urban architecture. Six 2008 winning Green Buildings projects, five in Manhattan and one in the Bronx, were selected for their exemplary integration of design and sustainability. The winners included Bowery Hotel and West Harlem Environmental Action, Inc.'s centre to be built at 459 West 140th Street. Hearst Tower and a condominium building at 1,347 Bristow Street in the Bronx, designed by the Community Environmental Centre, were the year's honourable mentions.

The eco-district models are multiplying in many cities (Arene 2007). New urban neighbourhoods offer ample opportunities for innovative resource, water and waste management, compact land use well integrated with public transport, cycling and transport and social innovations to improve lifestyles. The shared aim of many new developments is to offer citizens a high-quality life, within the limits of the fair share of Earth's resources. Developing lifestyles according to this philosophy and reducing environmental footprints is essential for residents of these developments. According to Aristotle, "habits are being formed from the first day."

The design and construction of Kronsberg, a new quarter of 70 ha, built in Hanover for Expo 2000, incorporated the available know-how on ecological aspects of construction and management, from energy-saving construction methods to sustainable rainwater and soil management and the exemplary waste concept. The model includeed the building waste concept, the commercial and domestic waste idea and communication. Within the construction waste concept, the city of Hanover reached a contractual agreement with developers to use exclusively ecological and healthy building materials. The amount of construction waste was reduced by 80% through prevention, sorting and recycling measures. Domestic waste at Kronsberg could be reduced to 154 kg per resident per year, compared with the Hanover average of 219 kg per person per year (2005).

The Beddington Zero Energy Development (BedZED) is the UK largest mixed-use sustainable community, designed to create a thriving neighbourhood with ordinary citizens enjoying a high quality of eco-responsible life. Highlighted in Suton's Agenda 21 and completed in 2002, the community comprises offices, commercial space, public and community areas and owner-occupied, privately rented and social housing units.

The holistic eco-design includes optimal conditions for resource and water management and incites citizens to reduce their environmental impact and make choices conducive to sustainable development. Electricity consumption is 25% lower than the UK average, the energy for heating has been reduced by 88%, water consumption by 50% and the distance travelled by private car decreased by 65%. The number of parking spaces is half the UK average of 1.2 spaces per housing unit.

Local authorities are trying out new types of district management, insisting on the quality of the daily environment. In the Netherlands, the management of the Romolenporder district in the Harlem community is a good example of ecological management. Inhabitants took part in planning and developing the district, constructing the wooden houses, with green roofs and clean energy systems. "Life on water" was the slogan used for the promotion of the residential area of Allermöhe in Hamburg, to welcome families and children to a high-quality environment which uses resources efficiently and respects all citizens, pedestrians and cyclists.

An interesting eco-neighbourhood was proposed by Stockholm on formerly abandoned waterfront industrial land, on the outskirts of the inner city, set aside for the city's ultimately unsuccessful 2004 Olympic bid. Hammarby Sjöstad is a new district with 10,000 flats, 10,000 jobs, cultural activities, services and parks with bioclimatic design and ecologically sound surroundings. The "eco-cycle model of Hammarby" aims to make the district autonomous throughout its life cycle. The fundamental principles include the decontamination of the soil, the optimisation of energy and water use and waste production, efficient public transport services, use of healthy and intelligent construction material and the keeping the noise level at 45 dB.

The "environmental impact profile" allows all the impacts on the environment to be calculated. Street rain water is collected, purified in a sand filter and released into the nearby lake, instead of draining into the sewage system and causing further pressure on the treatment plant. Stonecrop and sedum plants roofs help absorb rain water as do footpaths and trees and plants. The waste water from a single household produces sufficient biogas for the household's gas cooker. Most of the biogas is used as fuel in eco-friendly cars and buses. Intended to welcome 30,000 people by 2015, the district aims at a consumption of 50 kWh/m^2/year, instead of the current consumption of 120 kWh/m^2/year.

The city of Malmö, in noble emulation of the Swedish capital, proposed Västra Hamnen, a carbon-neutral residential district. It is composed of 1,000 low-energy-use buildings in a dense urban fabric served by sustainable transport. All energy needs for its 1,000 homes are satisfied by solar, wind and geothermal energy, from underground or seawater. At certain times during the year, the district borrows energy from the city systems, whereas at times, it supplies the city systems with its surplus. The overall balance is zero (Malmö 2008).

Three eco-cities have been promised by the UK Government by 2020. Many citizens rejected these plans, suggesting that they imply the construction of new highways to link them to the national network and demanded job generation along with the sustainable environments. Other voices against the government's plans suggest that the rehabilitation of the existing building stock, the development of the brownfields and the occupation of the 800,000 empty buildings are of higher importance than the creation of new urban centres.

The creation of a sustainable mixed quarter, surrounding and incorporating the EU headquarters in Brussels, has been the object of an international competition, part of the master plan adopted by the Brussels regional government in 2008. Brussels was long the provisional seat of the European institutions before becoming the official seat with the entry into force of the Treaty of Amsterdam (1997). The European quarter became the most important service district of the Belgian capital with 3.4 million square meters of office space; 1.9 million square metres of office space, spread over 55 buildings is occupied by the European Commission and the associated bodies. Keeping a strong symbolic presence is essential for the EU, while developing new sites to host the new officials.

Reducing the ecological footprint of the European institutions has been a strong cross-cutting objective and the project selected proposed a new urban model with a harmonious integration of spaces and functions. Reducing the emissions of the buildings, optimising the links to public transport and improving the management of spaces are the main principles. The project foresees the organic integration of the buildings in a genuinely mixed neighbourhood, with interacting commercial, residential and cultural functions to help bridge local, European and international universes.

In 2009, the European Commission adopted a guide on its architectural policy, which marks a major step as it gives a clear indication of its intentions to lead by example and significantly improve the quality of its buildings. The guide also includes the commitment to apply the Eco-Management and Audit Scheme to all European Commission buildings and organise architectural competitions for all new buildings and large-scale renovations to existing buildings.

Watercolour 3
Copenhagen: The Climate Capital

Copenhagen: The climate capital

Chapter 3
Transport for Sustainable Cities

Transport is a network industry and service and an engine and catalyst of growth and prosperity. It has vital links with all dimensions of sustainable development, but it experiences problems to be on a sustainability path. This chapter offers an insight into urban mobility and accessibility patterns and models and the challenges for profound change. The car is still by far the dominant transport mode in cities even though its supremacy is being challenged and citizens are asking for better transport services. Urban arteries often get blocked by traffic, pollution and emissions and may lead to urban thrombosis and asphyxia if a better balance between individual and collective transport is not urgently struck.

However, cleaner transport options and services exist and can be attractive. Clean and highly eco-performing public transport can be the backbone of sustainable urban transport services. Walking and cycling are highlighted as the only truly sustainable transport means. Car-free districts and cities should be rooted in the local culture and respond to the citizens' needs and concerns. Mobility weeks and symbolic events can boost the demand for competitive, attractive and environmentally friendly urban transport.

3.1 Mobility and Accessibility: Patterns and Effects

Transport is an engine of growth, jobs and prosperity. In the EU, it generates 7% of GDP and 5% of total employment. It is a network industry and service and offers mobility, closely linked to personal freedom and access, equality of which is a key social issue. Transport has vital interactions with all dimensions of sustainable development, socioeconomic, environmental and political, and can make a high contribution. However, the 2009 Review of the EU Sustainable Development Strategy signalled that transport in Europe is not on a sustainable path (EC 2009c, d).

In analysing the impact of the crisis on transport, the International Transport Forum suggested that the restoration of some global balances may reduce the intensity of the growth of trade, which, in turn, may affect future transport volumes (ITF 2009). Concerning urban traffic volumes, one should remember that the Asian

V.P. Mega, *Sustainable Cities for the Third Millennium: The Odyssey of Urban Excellence*, DOI 10.1007/978-1-4419-6037-5_3,
© Springer Science+Business Media, LLC 2010

crisis of 1997 had an impact on urban traffic in large urban agglomerations such as Seoul and Bangkok.

Transport infrastructures provide the arteries for the socioeconomic vitality to circulate to the last nerve of the city. They occupy 10–15% of the urban space and are the source of urban tensions and problems. Public transport (bus and rail) in the EU has been identified as one of the sectors where consumer satisfaction is the lowest (EC 2007c). Family spending on transport amounts to approximately 13% of total expenses. On average, an EU citizen makes 1,000 trips per year and half of these are less than 5 km long.

Mobility has long been regarded as a cardinal social value. The private car became a supreme symbol of individual freedom. Mobility patterns depend on both the supply of transport infrastructure and the increasingly complex and unsystematic demand. Demand is impacted by the largely unconstrained location decisions of firms, service providers and households. The interaction between transport and land planning of expanding cities is evident. Suburbanisation has always been inextricably linked to transport infrastructure. Many European metropolitan areas have suffered from a vicious circle of road construction and further suburbanisation (EEA 2006).

Better mobility and accessibility with less impact on the environment is a major challenge for European cities. Eco-innovation and eco-efficiency are major instruments. The distinction between mobility and accessibility is an interesting one (ALFOZ 1995). Unlike sheer mobility, accessibility is closely linked to services that citizens wish to have, even without moving. Accessibility is particularly crucial for people with reduced mobility and the elderly. The "Seniorcité" programme of the public transport enterprise in Paris is an interesting example of projects designed to respond to the accessibility needs of senior citizens.

Transport has witnessed important improvements in eco-efficiency over the last few years. However, increased transport demand has largely outstripped gains from technological improvement and fuel efficiency. Vehicle occupancies have declined and the vehicle stock and the number of kilometres travelled have grown far more than the population or GDP. Inland transport costs, in terms of accidents and environmental impact, are estimated to be as high as about 5% of GDP.

The transport sector had the highest growth of greenhouse gas emissions in recent years. In 2005, greenhouse gas emissions from transport accounted for 21% of total EU emissions. The great bulk of them came from road transport (93%), still depending almost totally (97%) on fossil fuels. At the global level, road transport has become the largest single and most intractable anthropogenic source of CO_2 emissions. Addressing greenhouse gas emissions from transport requires technological, social and political innovation, demand-side management and governance, with the public engaged in policy development and debate (EC 2008e).

Private cars represent the dominant urban transport mode. Economic growth has been a major factor for increased car ownership and transport flows. Congestion in European cities often makes average traffic speeds at peak times lower than in the days of the horse-drawn carriage. Increased car use has been the cause of safety and environmental problems, as well as of a downward spiral of underinvestment in public transport. One third of fatal road accidents happen in cities. Particulate matter and noise levels from transport are also worrying. Substantial progress has been made over the last decade in reducing vehicle emissions, but growing traffic levels underpin progress.

Decoupling economic prosperity from emission and pollutant release from the transport sector is an important EU objective. The 2009 Review of the EU Sustainable Development Strategy revealed that, although GDP growth has exceeded growth in energy consumption by the transport sector, energy consumption and greenhouse gas emissions from transport continue to rise. Despite the signals of decoupling transport volumes from economic growth in passenger transport, freight transport has grown faster than GDP, since the previous 2007 review. Emissions originating from passenger transport increased by an average of 1.7% per year during the period 1995–2007, whereas emissions from freight transport increased by an average of 2.7% per year during the same period, against an average 2.5% increase of GDP per year. Air pollution and noise caused by transport and also road accidents remain major concerns for Europeans (EC 2009c, d).

Physical accessibility and mobility can be optimised through planning measures and economic instruments. The elimination of architectonic barriers, in particular relics of heavy transport infrastructure fragmenting many urban landscapes, and the designation of the recovered space for public use can lead to fluid cities which facilitate the mobility of all. Virtual accessibility is also important and can complement physical measures to ensure fluid movement and access.

Freight transport responds to the right of access to markets and goods and accounts for about 10–12% of vehicle traffic in cities but causes disproportionately higher impacts on congestion and the environment. Freight is essential for the economic functioning of cities, but the efficiency of the sector has much potential for improvement. Construction works and retail are responsible for much of the goods traffic. Commercial vehicles vary largely from dirty, noisy and intrusive trucks to smart electric cars. Many cities impose vehicle size or weight restrictions, or limit access in certain areas. A balance has to be struck between access requirements and transport and environmental objectives. Policy responses include the German model, based on private transport companies coming together to serve, and the Dutch model, based on licences provided by authorities to transport companies.

Urban policy portfolios to optimise transport include the promotion of lower-consumption vehicles and new propulsion technologies, demand-management schemes, such as parking and access restrictions, pricing regimes and land-use planning, information and communication. Reconciling and mixing land uses is essential for reducing transport flows. Population density may be inversely related to travel demand and energy consumption. Optimal interrelated transport and land-use interventions depend highly on the local context. Many parameters have to be taken into account, including the size and structure of the city, the location of the various activities, the distribution of the employment, public transport alternatives and the parking policy (OECD-ECMT 1994).

At the EU level, transport forecasting scenarios consider external drivers (population, economic development, energy, technology and social change), internal drivers (infrastructure, vehicle and fuel development and socioeconomic impacts) and policy drivers linked to the governance of the transport sector. Exploratory models ("Move alone", "Move together", "Move less" and "Stop moving") and metamodels based on a set of interdependencies (between external input and resulting output) gave interesting insight. An anthropological constant, expected to continue in the future, is the average commuting time (about 1 h) per person per day (Tetraplan 2009).

The sustainable development of European cities can benefit much from EU actions on urban mobility and accessibility, with special attention to the needs of vulnerable groups such as children, low-income groups and persons with disabilities and reduced mobility, who, all together, account for one quarter of the total EU population.

In 2007, the Green Paper on Urban Mobility identified five main challenges: fluid urban transport, greener towns and cities, smarter urban transport, accessible urban transport and safe and secure urban transport. The consultation and debate that followed the publication of the green paper confirmed the importance of these issues for European citizens and highlighted the benefit of action at EU level. Best practices, indicators and benchmarking are amongst the higher-value European initiatives (EC 2007c).

European programmes of research, technology development, demonstration and innovation have already added value to the urban innovation chain. The European CIVITAS (City–Vitality–Sustainability) Initiative, launched in 2000, generated a decisive breakthrough by supporting the implementation of ambitious integrated sustainable urban transport strategies. The participating EU cities have made exemplary efforts in developing attractive alternatives to the use of private cars, through an integrated set of technology and policy measures to make a real difference for European citizens. CIVITAS I (2002–2006) brought together 19 cities clustered in four demonstration projects and CIVITAS II (2005–2009) brought together 17 cities in four demonstration projects. CIVITAS is a programme "of cities for cities", living "laboratories for learning and designing urban futures". The CIVITAS annual forum brings together experts and politicians to exchange ideas and practices.

3.2   High-Quality Urban Public Transport

It seems that the transport system of the future will be far more diverse and less dependent on cars. Light rail, buses and bicycles may be much more important features of the transport services (Brown 2006). Public transport can only be an attractive alternative to the car if it is safe, clean, reliable, fast, frequent, noiseless, flexible, easily accessible, well designed, environmentally friendly and economically viable. It plays a major role in the larger cities, where it carries 2.5–3 times as many people as private transport. Public transport is also important for an estimated 40% of EU households that do not have a car (EC 2007c).

Highly and densely populated cities such as Tokyo, where only 1% of commuters use their private car, can provide many inspiring lessons in upgrading public transport systems. In Europe, Zurich and Basle have the highest public transport use. Zurich has developed an eco-performing approach for a public surface system. Preserving and upgrading the tram system and rearranging the bus lines were the key elements in the improvement of the public network. The system achieved high speed by allowing each public transport vehicle to cross intersections without stopping.

The Zurich public surface system is received very positively by the population, which has been almost unanimous in promoting its expansion. A high number of seasonal tickets are issued for the system owing to the support of enterprises which offer free or reduced tariffs to their employees (EEA 2007a).

Innovation in public transport spans a multitude of issues, from the fuel source and system design to ticketing and demand-management measures. Increasing use of hybrid buses, electric buses, biobuses and hydrogen buses and intelligent transport management systems is developing continuously. Lille developed a fleet of buses running on biogas, whereas in Graz all buses use biodiesel. Green lanes exclusively reserved for public transport are a common feature in many cities. Heidelberg, Freiburg and Basle were pioneers in introducing low-noise vehicles in noise-protection districts. Increasing use of smart transport passes, often linking public transport to other social and cultural services, and eco-tickets is progressing continuously (UITP 2007).

All urban public transport systems in the world are becoming smarter and are integrating intelligent systems to offer new services. The metropolitan underground systems are going through major innovations, such as driverless operation, high-tech information, ticketless services and security devices. Lausanne features a fully automatic metro line, whereas front-runners such as Dubai, Hamburg, Montreal, Paris, New York, Tokyo, Berlin, Moscow, Barcelona and Singapore have important lessons to share.

Light-rail trains have become a familiar feature in many cities, whereas tramways, abolished in many European cities during the 1960s, made an impressive come back in the 1980s. Nantes, Grenoble and Strasbourg introduced from 1985 onwards three technological generations of tramway. Valencia advertised its tramway as a "tramway named desire". The tram, abolished in 1960 in Athens, made its return in 2004, just in time for the Olympic Games. Equipped with the latest technology, it offered the city the possibility of a valuable alternative, complementary to the expanding bus and metro networks (ERRAC 2004).

Transport is a network industry and service. Pertinent articulation among bus, boat, tram, light-rail train and underground systems can make a difference. In Brussels, the network strives to combine the efficiencies of metro, pre-metro and tramway systems. The tram–train system of Saarbrücken shares existing tracks with the national railway, making infrastructures more efficient. The greater Stuttgart region, responsible for transport planning and investment, introduced an integrated traffic and transport concept with buses becoming parts of the extension of the suburban electric railway network. Effective national–municipal cooperation between the four municipalities of the Helsinki metropolitan region and the railways has been instrumental in keeping passenger cars outside the urban core. Integration of urban, suburban and short-distance regional transport services is traditionally covered by public service contracts and operators have the obligation to safeguard passenger rights and quality of service.

In historic Italian cities, various alternative-mobility schemes have addressed the need to improve urban quality of life threatened by tourist pressure often concentrated in space and time. In Orvieto, tourist buses were driving to the hilly historic town, causing much damage to the fragile rocky morphology. The city revitalised the old funicular railway and introduced a car parking system with parking spaces at the foot of the Orvieto hill. All visitors can take the funicular railway to the top of the hill and then use the network of minibuses in the city. A single authority for public transport and private car parking helped internalise more equitably the environmental costs of private motoring and the improvement costs of public transport.

Best practice in intermodal travel information, access control and demand management, more accurate positioning of vehicles, traffic regulation, electronic ticketing and payment and partnership approaches are cardinal. Budapest introduced a complete ticketing integration of public transport providers, including the city's local public transport company, the national railway and the long-distance coach company. A combined monthly pass allows the use of any service within the metropolitan boundaries.

Adapting supply to fluctuating demand patterns has given rise to many responses in Europe. In an early scheme, Perugia introduced in 1985 a telebus service, especially for peripheral zones, running along a principal route, and able to serve additional secondary routes only upon request. The service was enabled by a communication centre and personal magnetic cards of the users. The system has proved very efficient and it is particularly inspiring for areas with sparse settlements and for citizens with reduced mobility.

The sociocultural aspects of the public transport systems deserve much attention. Transport places are spaces of human convergence, spaces of passage and of sharing, of exchanges and communication. Railway and coach stations, as airports, are places to celebrate the act of departure from and arrival at a city. Many transport sites serve as exhibition and cultural places, a prelude to a city (UITP 2005). Major stations of the new metropolitan system in Athens host archaeological galleries that also exhibit the findings during the excavations for the underground.

The image of public transport has to be improved in many cities of the world. In Budapest, the administration of the metro, around 100 years old, is still seen as embodying arbitrary features of the old regime. A young cineaste devoted a film to the metro controllers. The film was paid for by the administration, which was criticised as a result.

Worth noting examples of public transport also come from "asphyxiated" world cities, especially in the developing world. Bangkok, one of the world cities experiencing the most difficult traffic conditions, has also made important steps over the last 10 years. The aerial train Bangkok Mass Transit System, inaugurated in 1999, added an important service to the congested transport landscape of the Thai capital. Each train has a capacity of 1,000 passengers and represents the equivalent of 800 cars. Quick, secure, agreeable, environmentally friendly and citizen-friendly, the skytrain has been embraced by inhabitants and tourists. Although the skytrain option seduces cities with space scarcity, some cities, such Sydney, went even further, with the skytrain crossing the built environment.

On the South American continent, when in the 1960s most Brazilian cities built highways to serve the growing private car fleets, Curitiba took another, environmentally sensitive path. It gave priority to its public transport system. Today it has a highly performing surface system, exclusively constituted of buses. The buses are long, biarticulated and especially designed for the city, have a capacity of 270 passengers and stop at designated elevated areas, equipped with access for the disabled. The system is integrated, with a unique fare, paid at the bus stops. The result is that 85% of the population use public transport (UNDP et al. 1993).

3.3 Walking and Cycling in Civilised Cities

Sustainable development asks for citizens to adopt less-car-dependent lifestyles and explore new forms of mobility, for example in the form of walking, cycling, car sharing, car pooling, bike sharing and taxi sharing. The human leg is the only truly sustainable transport means. Walking and cycling have gained ground in most European cities (EEA 2009c).

The human face of pedestrian-friendly cities has often been praised. The transformation of central districts into pedestrian areas has been a general trend in most European cities since the 1960s, despite strong opposition from commercial lobbies. Copenhagen has been a pioneer city in recognising the social value and the economic benefits of central pedestrian streets. The creation of pedestrian precincts started in the 1960s and evolved in coordination with the downtown parking policy and the elimination of 2–3% of the parking spaces per year. The public transport system was improved and the bicycle network enlarged. More and more space was liberated from traffic and rendered to citizens, who started leaving behind anonymous peripheries and coming back to the city centre. The pedestrian Stroget area attracted civic architecture, sculptures, fountains and musical and cultural events.

The role of the street as a shared public space is dignified in the European Urban Charter issued by the Council of Europe in 1992 and reinvigorated in 2008 (CE 1992, 2008). The "Code of the street: streets for all", introduced in 2004 in Belgium, asks for more attention to be paid to the rights of the pedestrians, cyclists, children, the elderly and the handicapped. The code requires drivers to respect a 1-m minimum distance from pedestrians when the latter are allowed to cross streets. It also reinforces repression measures. The concept of crossing curbs can promote safe crossing and incite drivers to reduce speed.

Bicycles are the only other sustainable transport means, second only to walking. Policy measures and infrastructures to promote cycling are expanding in most European cities. The modal split of cycling has grown significantly in European cities during recent years (EEA 2009c). Amsterdam and Copenhagen are capital cities endowed with the most elaborate bicycle network, complementing the road and canal routes. Both capitals developed successful public "City Bikes" programmes.

Copenhagen, selected by the International Cycling Union as the world's first Bike City, registered 29% of commuting trips by bicycle. Like Hanover, Copenhagen implemented speed reduction on a significant proportion of its roads (EC 2007f).

Creating the necessary infrastructure and creating the conditions are fundamental policy measures to support cycling. Helsinki has created more than 1,500 km of bicycle lines, whereas Stockholm and Hanover have created about 750 km. Vienna and Munich have reported a high number of trips by bicycle for commuting. Smaller cities (fewer than 300,000 inhabitants) such as Turku, Aalborg, Tampere and Aarhus, offering a cycling network of about 300 km, managed to convince about 20% of their citizens to use the bicycle also for home-to-work trips.

Cities such as Basle can be crossed and enjoyed by bicycle. Since the early 1990s, Zurich and La Rochelle have lent bicycles free of charge to residents and visitors. The Velib service, involving flexible bicycle renting (bike-sharing) through an electronic network, has been successful in Lille and Paris, as has the equivalent Villo service in Brussels. Brussels also organises the Velo-city event to promote pedal power.

The 15th Velo-city event, in 2009, focused on "recycling cities". Four plenary sessions and 28 subsessions concluded with mayors of many cities signing a cycling charter at the European Parliament. The hosting city is about to renew its very own bike-sharing system with 180 stations and 2,500 bikes. Cycling is also promoted with campaigns such as Friday Bikeday and the annual bike festival DringDring, an annual cycling-promoting event for all and a night cycling event.

A broad range of EU funding initiatives encourage cycling as an urban transport mode. They include investment in cycling infrastructure in eligible regions across the EU, the implementation and evaluation of 35 cycling-related measures across the cities that participate in CIVITAS and provided €10 million to ten European pilot projects related to cycling through the STEER Programme, which promotes more sustainable energy use in transport. Programmes such as CIVITAS and STEER have helped local authorities to turn bicycles into a viable alternative to motor cars. A cycling training for school children in London doubled the number of children cycling to school at least once a week. Gothenburg promoted cycling by doubling the number of cycle parking spaces from 7,000 to 14,000. Lyon saw the number of people cycling go up dramatically after it introduced a bike-sharing system and improved the infrastructure.

In Leipzig, local authorities encourage the inhabitants to use their bicycles to connect to the public transport network, to which they are offered access for free if they park their bicycle in a bicycle parking space. This "Bike and Ride" system is enhancing the 200-km bicycle network, which offers the basis for 13% of all urban trips.

3.4   Cleaner Transport Concepts and Technologies

Almost all (97%) of energy in the transport sector is produced from oil-derived products (2005). Biofuels, originating from organic, if possible waste, matter are the main substitutes for petrol and diesel, but account for only a minimal proportion of total EU energy consumption of transport fuel (2.4% in 2006) . Their use may increase much, given that, within the ambitious EU target for renewables to reach 20% of the energy mix by 2020, fuels for renewable sources have to account for 10% of the vehicle fuel. Many European cities provide an example with parts of their municipal fleets running on alternative motor fuel.

Most European sustainable transport approaches aim at making passenger cars more environmentally friendly and achieving a significant switch from the use of private cars to public transport, cycling and walking in cities and their commuter belts. In 2009, the EU adopted a regulation setting binding targets for CO_2 emissions from new passenger cars and a directive for the promotion of clean and energy-efficient road transport vehicles. Complementary policies include economic measures, such as increasing taxation to make company cars less attractive, taxation of free off-street parking spaces and road pricing. By making users pay the full environmental, public health and congestion costs, internalisation of external costs can encourage transport users to switch over time to cleaner vehicles or transport modes, to use less congested infrastructure or to travel at different times.

The European JUPITER and other targeted transport projects demonstrated the potential of innovative energy and environmental technologies. The JUPITER II project achieved reductions in energy consumption of 20% and in emissions of harmful air pollutants of between 16 and 25%. The strategic assessment of the project highlights that the modal split has significantly improved in most cities, with a 12% increase in public transport. A more extensive implementation could double this figure. A substantial reduction of 4% in CO_2 emissions and 20% in particulate emissions has also been recorded.

The EC-supported project ZEUS (zero- and low-emission vehicles in urban society) is led by Stockholm, in collaboration with other European cities. One of the ZEUS subprojects focused on the introduction of vehicles fuelled by biogas, originating from recycled liquid organic waste. A pilot station for biogas production has been constructed and hybrid vehicles functioning with petrol and/or biogas were gradually introduced. A biogas-fuelled lorry transports biogas to the filling station. A fleet of 200 vehicles, lorries and private cars has been the result of the cooperation among the municipal enterprise for wastewater management, fuel companies and city infrastructure services.

Intensive EU-supported research on fuel cells and hydrogen is progressing. Initiated in 2001, the CUTE project is regarded as a most important demonstration project on hydrogen. Hydrogen-fuel-cell buses have been operating in regular public transport services without noise or other nuisances. Other hydrogen demonstration projects include large projects developing technologies for cars, buses and small transport vehicles as well as exploring infrastructure solutions for the supply of hydrogen.

Cleaner public transport fleets for cities could help reduce fuel consumption as well as CO_2 and pollutant emissions from road vehicles. A substantial reduction may be achieved by public authorities procuring clean and energy-efficient vehicles for public transport services. The introduction of environmental criteria into the requirements for public procurement of vehicles by public authorities and by operators providing public transport services could lead to substantial improvements. When planning the purchase of vehicles, public authorities could opt for joint procurement of clean and energy-efficient vehicles for public services. Award criteria should use lifetime costs for CO_2 and pollutant emissions as well as fuel consumption.

Social innovation to decrease the dependence on private car use is very important. Concepts proposed by cities or businesses to pool commuters can be very beneficial. Modena experienced a novel bus-pooling experience through the joint action of two companies introducing a free bus service for commuting by their staff.

The careful articulation of transport means offers new possibilities for partnerships. The AUTOPLUS system in La Rochelle created new synergies between bus companies, taxis, boats and hotels. The park-and-ride project "De Slinge" in Rotterdam transformed a former car park from an arena of vandalism into a welcoming public multipurpose station, where citizens often exchange their cars for their bicycles.

The Pegasus award promoted by Suisse Energie and Sun21 for the most promising projects of sustainable mobility allowed identification of projects coupling technology with social innovation. Access for all to the mobility car-sharing park, multitalented bicycles convertible into caddies, imaginative solutions to reduce the number of shopping-related trips, enterprise mobility plans and a virtual school of environmental management constitute a bouquet of highlighted actions to improve urban mobility.

3.5   Car-Free Cities?

"Is the city without car" a Utopia or a reality of tomorrow? There is consensus that a city cannot exist without movement, but many challenge the possibility of a modern city existing without cars. Is there an optimal volume of traffic that a city can benefit from? The efforts of most cities to limit the numbers of cars in specific urban districts indicates that a critical boundary has been crossed and problems threaten the benefits.

The Italian cities have been pioneers in closing their historic centres to private cars and introducing park-and-ride systems. Leading cities, such as Rome, Milan, Florence, Bologna and Bolzano, experimented with various scales of car restrictions in their historic centres, after an intense debate with citizens. Venice remains the archetype of a car-free city, but its unique physiognomy cannot indicate a model to be replicated. As in most cities, infrastructure pricing provoked a heated debate in Venice, with a hot discussion about the pricing of access to the car-free centre for tourists.

Perugia was one of the first cities to close its historic heart to private cars. Furthermore, it created a cultural passage linking the central pedestrian area to the parking spaces, including a gallery of urban archaeology. In Portugal, an inspiring model was proposed by the city of Evora, which closed its historic centre, surrounded by an ancient wall, to private cars, and introduced a high-quality public transport system, with minibuses and microbuses adapted to the narrow medieval streets, large car parks outside the city walls, park-and-ride sytems and agreeable pedestrian streets and cycle tracks (Mega 2008).

In Curitiba, the elimination of private cars from the city centre was part of a visionary plan to oxygenate the heart of the city. In 1972, on the initiative of an expert mayor, within 3 days and despite protests by the shop owners, several commercial streets were transformed into pedestrian ones. When detractors of the project tried to enter the pedestrian areas by car, the mayor gathered hundreds of children and gave them large paperboards to paint on the paving stones, a legendary activity which continues every Saturday mornings.

Infrastructure pricing can be an effective policy measure for limiting the number of cars in cities. Urban tolls usually meet opposition from car users and the car industry. Referenda in Edinburgh and Manchester rejected the scheme with 75% of the votes. The experiences of Oslo, Stockholm and London with urban tolls offer many lessons on citizen consultation processes, scheme design, information provision, public acceptance, operating costs and revenue, technological aspects and the impact on the environment.

As early as 1990 and following the example of Bergen, Oslo introduced road pricing, after 10 years of public debate. Nineteen stations were installed, 3–8 km from the city centre. Since the introduction of the toll ring, the number of cars entering the city centre has decreased by 3.5%, whereas there was an increase of 10% of trips by public transport crossing the toll ring (EEA 2005, 2007a).

The limitation of cars entering central London, introduced in 2003 and extended into parts of west London in 2007, aimed at reducing congestion and raising funds for investment in public transport. The congestion charging scheme represented an investment of £100 million in electronic equipment. The number of cars entering the City of London has considerably diminished and the annual revenues are estimated at 20% higher than the annual operating cost of the system.

The improvement of public transport is the prime condition for citizens to accept not using their private cars, should an attractive alternative option be offered. Urban tolls can become more sophisticated with improved price differentiation according to the time of the day, the length of the journey, the engine size and the cleanliness of the fuel. The potential of European short sea shipping should be overhauled. Among other schemes to limit the number of private cars, car-pooling and car-sharing schemes have expanded with mixed results and prospects. The Auto-lib service, flexible car renting in city centres, following the success of the Velib service in Lyon, Paris and Villo in Brussels, could also decrease the number of private cars entering cities.

Many German and Swiss cities have experimented with the creation of car-free residential neighbourhoods. In Bremen, seven families with children, living in various districts, were chosen to take part in an innovative experiment: stop using their car for 4 weeks and maintain a diary of the changes that this experiment would bring to their life. The analysis of the diaries revealed that nobody felt restricted in their daily movements, except for excursions at the weekend, outside the city. All the participants mentioned the role of their daily functioning in a social place (as opposed to the car, private space) and the sensorial perception of the surrounding space (Burwitz et al. 1991).

Research on "a city without cars" conducted by the European Commission in the early 1990s suggested that the city could be re-examined and be conceived for pedestrians. A city without cars could be made up of various small units on a human scale which would be connected by high-speed means of transport. It seems that the city without cars would not only be effective at the ecological level, but also efficient at the economic level, since the cost of managing more dense urban areas is estimated to be considerably lower (2–5 times depending on the density). An attempt to adapt the ideal model offered by this study to the real conditions of the European cities connects pertinent transport spaces with public spaces.

The attraction of these concepts led the municipality of Amsterdam, one of the cities which, like Bologna, had organised a referendum on the restriction of the use of private cars in the downtown area, to organise the conference "Cities Without Cars?" The question mark is significant, as it expresses reactions and inhibitions. On that occasion, cities committed to promoting policies discouraging the use of private cars launched the Car-Free Cities Club. The passage from car-occupied spaces into noble citizen spaces represents a major challenge for European cities (Municipality of Amsterdam 1994).

Symbols are powerful for raising awareness. Since 2002, the European Mobility Weeks have encouraged and celebrated the various alternatives to car driving: walking, cycling, public transport, car sharing, lift sharing, eco-driving, etc. A wide range of initiatives addressing different aspects of urban mobility are organised by local authorities on each day of the week and in partnership with local organisations and associations, NGOs and businesses. The Car Free Day is the highlight of the week, with the challenge of organising "In Town Without My Car" on a working day!

Each European Mobility Week represents a platform for local authorities as well as organisations and associations to promote their existing policies, initiatives and best practices on sustainable urban mobility, launch new policies and initiatives and raise citizens' awareness on the damage that urban mobility trends cause to the environment and the quality of life. Many events highlighted the importance of the safest possible mobility patterns for children.

In 2006, the European Mobility Week had the theme "The Street for All" and was celebrated in 1,322 cities. Many projects focused on reinventing the street as a noble shared space and experience. The theatre of the street has proved to be an excellent means of participation and of communion. In Brussels, the talking bus proposed "discovering, dreaming and liking Brussels", so that that inhabitants would appreciate their city.

Lisbon's participation in European Mobility Week 2007 included a "Zero Tolerance" campaign aimed at forbidding car parking on sidewalks or in a second row. "The Streets for Living" event included 20 main avenues, streets, squares and roundabouts all over the town. Furthermore, the week stimulated progress towards more permanent features affecting everyday life. The public transport network was improved with two lines of buses providing permanent access for disabled citizens. Two other lines allow bicycles accompanying passengers to be transported free of charge at weekends and on holidays.

The theme of European Mobility Week 2008 was "Clean Air for All", an issue which ranks high on the scale of citizens' concerns. In Cork, nominated as the best Irish participant, the range of events included an air quality awareness campaign, cycling workshops, free health screenings, movies, free bike training courses for families, a sustainable transport exhibition and a showcase of electric bicycles and scooters. The "Rebel Pedal!" bicycle parade and picnic were the highlights of the week, and there were several prizes, including a professional camera and a luxury weekend break. The Black Ash Park and Ride service remained car-free throughout the whole week and provided venues for the official launch of the new GoCar (CarSharing) and Avego (RideSharing) services.

The 2008 European Mobility Week Award was won by Budapest, judged by an independent panel of experts to have done the most to raise public awareness of air pollution from traffic and to promote cleaner alternatives during European Mobility Week. The city and districts of Budapest presented a rich range of activities to promote sustainable urban transport, with the involvement of local artists and organisations.

The activities included two consecutive car-free days, a "Clever Commuting Race" for VIPs to demonstrate the efficiency of public transport, an open-air exhibition of clean and energy-efficient vehicles, a conference on air quality and noise mapping and a "Pedestrian Areas Day", promoting walking in the historic city centre along the banks of the Danube. With this occasion, the Hungarian capital introduced and promoted some permanent measures for more sustainable transport, such as the expansion of the downtown pedestrian area, an increase of parking fees in the city centre, improvement of public services, and new bicycle lanes and park-and-ride facilities.

Almada, the runner-up, used the opportunity of European Mobility Week 2008 to convert its historic centre into a pedestrian area, launch three light rail lines, reallocate road space in favour of sustainable transport modes and introduce new bus shelters. In cooperation with the university, the city launched a permanent air quality monitoring system. The events organised included a "Public Transport Day" with live music, free coffee and presents for passengers, a "Commute by Bike Day", a "Mobility Management Day" and a "Local Shopping Day". During the car-free day on 22 September, the university area was closed to motorised traffic and a series of awareness-raising activities targeted students and school children.

Zagreb, the other finalist, introduced new trams and buses, improved public transport services and extended the infrastructure for cyclists and pedestrians. It also increased parking fees, introduced parking limitations and permanently reduced car traffic in the upper part of the city. A major public debate on air quality was held and a "Public Transport Day" was organised to promote the benefits of public transport and to make car users aware of the priority of trams and buses on the dedicated lanes. During the car-free day, 6 km of streets were transformed into pedestrian zones and a huge bicycle ride was held through the city centre, together with public discussions, lectures, sports events and exhibitions.

"Improving City Climates" was the theme of the European Mobility Week 2009. The year 2009 marked an especially crucial year for combating climate change.

As the eyes of the world were turning to the UN Climate Change Conference for a post-Kyoto agreement to be achieved in Copenhagen, cities presented their plans to reduce their carbon footprints of urban traffic. Although the latest technological developments regarding clean and energy-efficient means of transport, alternative fuels as well as intelligent transportation systems look promising, the effects of global warming can only be countered by making the change towards more sustainable transport modes such as cycling, walking, public transport, car sharing and car pooling.

Watercolour 4
Brussels: The Pedestrian City Centre

Chapter 4
Energy for Sustainable Cities

Cities and governments have to ensure a secure, clean, competitive and affordable energy supply. They also have a responsibility to influence demand patterns linked to consumer behaviour and society lifestyles. Forward-looking studies have highlighted that, in the absence of bolder policies, the energy future will be unsustainable in all its dimensions, environmental, economic and social. But this can and should be altered. The 20%/20%/20% European targets for 2020 can be mobilising for cities, which often have more ambitious plans to improve energy efficiency, reduce greenhouse gas emissions and increase the share of renewable energies in their energy portfolio.

This chapter examines renewable energy sources and cleaner energy options and technologies and highlights worthy efforts in cities to realise the potential of solar and wind energy, bioenergy, fuel cells and hydrogen and to create distributed networks. It presents ethical questions linked to access and use of energy and reviews citizen perceptions and concerns, addressed by a recent European energy poll. It finally argues that nuclear fusion may be a competitor for renewable energies in some decades from now.

4.1 Secure, Clean, Competitive and Affordable Energy

Sustainable energy policy portfolios aim at achieving a meaningful balance between security of energy supply, competitive energy services, affordable prices and climate protection. Cities and governments have to ensure a secure, clean, competitive and affordable energy supply and respond to contingencies, such as disruption of electricity or oil supply, quickly and effectively. Public policy has to optimise energy supply and use while minimising its impact on the environment, integrate the complexities of the evolving energy systems, establish targets, evaluate progress and adjust measures to meet the objectives (IEA 2003a, 2008a).

Energy policy options depend on geopolitics, international trade, market liberalisation, environmental concerns and concerted actions against climate change. The EU is the largest energy importer in the world. In 2006, 54% of the energy consumed in the EU was imported. Almost all member States rely on energy

V.P. Mega, *Sustainable Cities for the Third Millennium: The Odyssey*
of Urban Excellence, DOI 10.1007/978-1-4419-6037-5_4,
© Springer Science+Business Media, LLC 2010

imports to satisfy their gross inland demand. Energy demand increasingly outstrips indigenous production and external dependence is rising. Energy dependence can have serious consequences, such as supply uncertainty, higher energy prices and exposure to political instability of exporting world regions.

Promoting competitive energy and creating new business opportunities in the related industries and services constitute prime EU policy objectives. Government involvement in energy was traditionally seen as a national security issue. Over the past few years, virtually all EU member States have introduced competition in the electricity and gas sectors. Market options were multiplied and rising competition unlocked new opportunities. Electricity markets gradually converged towards a common approach, allowing consumers to choose their supplier.

Denmark is the only energy-import-independent EU country, whereas some countries, such as the UK and Poland, have low dependency rates. Small countries such as Luxembourg, Cyprus and Malta are completely energy dependent, whereas Spain, Portugal, Ireland and Italy have import-dependency ratios exceeding 80%. The structural weakness of the EU to balance its energy system can be witnessed in all sectors of the economy. Transport and the residential sector largely depend on oil and gas and are at the mercy of unpredictable globalised markets (EC 2008b).

At the start of the millennium, the European Commission (EC) green paper on a European strategy for the security of energy supply generated a vigorous debate in search of meaningful policy measures. Industrial production patterns have largely been adapted to environmental requirements. They are more concentrated in space and easier to target and change. The green paper advocated a real change in consumer and society behaviour. Taxation measures can steer demand towards energy options that are more respectful of the environment and generate revenues to be invested in environmental protection (EC 2000).

A reliable, flexible and diverse energy supply is necessary for cities to satisfy the needs of their citizens. No single energy option has the capacity to fulfil all energy needs in the immediate and near future. Diversification is necessary and has to be reflected in political priorities. Science and innovation have an essential role to play in improving energy efficiency and in exploring and capitalising on the potential of all energy.

The EU Energy Policy, launched in 2007, proposed a vision of full decarbonisation for 2050 and set a number of ambitious targets for 2020: reducing greenhouse gas emissions by 20%, improving energy efficiency by 20%, raising the share of renewable energy to 20% of the energy mix and increasing the level of transport fuels from renewable sources to 10%. All member States prepared action plans with targets for emission reductions and the share of renewable energies and cities committed to achieving often more ambitious targets (EC 2007a).

Cities have to ensure that energy systems are economically robust, socially beneficial and environmentally sound and that energy services meet the expectations of citizens. Local governments play an important role in leading communities into the postcarbon age with renewable energy, energy efficiency and energy savings as major cornerstones. A bold vision for cities could be to become net producers instead of net consumers of energy. Urban buildings and districts could be converted into active energy generators.

During the 2008 European Sustainable Energy Week, a major awareness-raising event, the EC launched a most ambitious initiative involving jointly cities and citizens in the energy future. Nearly 100 mayors from across Europe signed up to the Covenant of Mayors, a prime commitment to go beyond the EU targets. With 700 signatories and 33 supporting structures in 2009, the initiative is ambitious and result-oriented and brings together the most pioneering European cities to exchange good practices on energy efficiency and promote low-carbon business and economic development. Commitments include the submission of a Sustainable Energy Action Plan within 1 year following the formal signing up to the Covenant of Mayors, periodic monitoring, reporting and benchmarks of excellence.

Under the umbrella of Énergie-Cités, 1,000 cities, presently led by Heidelberg, invest in action to improve energy performance and generate energy savings (Énergie-Cités 2001). But cities, although concentrated major energy-demand centres, are not unrelated to energy and electricity production, even though this production takes place away from the cities and is managed by different governance structures. They have to participate in all decisions on the future of energy production, together with the other levels of governance.

Energy consumption in the EU is mainly fuelled by fossil fuels, which still form the backbone of the primary sources of world energy. Although the reserves of fossil fuels are finite, a global crunch is less feared today than a few decades ago. However, their use has serious consequences for the environment and climate change. Experts and politicians suggest that, as the Stone Age did not come to an end because the world ran out of stone, the fossil fuel age will not come to an end with the depletion of the deposits, but with human ability to move to higher-value fuels (Mega 2005).

Eighty percent of the fossil fuel reserves of the EU are solid fuels, including coal, lignite, peat and oil shale. Along with steel, coal was regarded as a cornerstone of the economies in the early years of European construction. The primary objective of the European Coal and Steel Community Treaty, in 1955, was to establish a common market in coal and steel and to contribute to economic growth and employment generation. Demand soon outstripped supply and greater production was encouraged.

Coal mining, a labour-intensive industry, contributed considerably to the full employment economy of coal regions after the Second World War. However, after the 1960s, coal mining went into rapid decline owing to international competition and the demand for cleaner energy sources. Lignite and peat production are profitable businesses, but the average cost of coal production, owing to geological and labour conditions, became much higher than the international market price. Since 1990, imports, mainly from the USA, Australia, South Africa and Columbia, have exceeded the indigenous production. In 2000, the EU produced around 5% of the world output (EC 2000).

The decline of the coal industry became the epicentre of political turmoil in some countries, in particular Germany. The coal compromise, concluded in 1997 between the German Federal Government, the Länder and the private sector, led to a considerable reduction in state aid, production and employment. Some EU member

States (Portugal, Belgium and France) decided to cease all production. Germany and Spain opted to restructure the industry and the UK decided to make production competitive with that of imported coal. Ensuring access to certain reserves and minimal capacity of production in realistic economic conditions could be instrumental for the continuity of operations and for allowing European technology to keep its leading position in clean coal production.

Among the fossil fuels, oil, used in transport and for electricity and heating, has the lion's share in the primary energy mix consumed in the EU (37% in 2006). Oil reserves are very unevenly distributed around the globe. The EU produces scarcely 4.4% of the world output. Resources are mainly located in the North Sea and belong primarily to the UK. Oil makes up the bulk of total EU energy imports, most of it coming from OPEC and Russia, followed by Norway and Kazakhstan. The cost of extracting one barrel of oil in Europe is around 4–6 times more expensive than in the Middle East (EC 2008b).

The natural gas discovered in the 1950s and once considered as a secondary by-product of oil exploitation has become a polyvalent source of energy. Cleaner and easy to use, with its own distribution network, it has gained ground in all sectors, including power generation, production of heat and transport. The EU produces 12% of the world output. Most of these reserves are located in the Netherlands (56%) and the UK (24%). Import dependency has increased more rapidly than for any other fuel.

The extraction, processing, distribution and use of fossil fuels have important environmental implications. During the extraction phase, methane and carbon dioxide may be released, for example when natural gas, which is typically 85–95% methane, is flared from oil wells. Oil and natural gas usually occur together in deposits and oil drillers may burn off the gas or release it straight into the atmosphere. Natural gas may also be released during the extraction and processing of coal. As the coal is being extracted, open gas pockets inadvertently break, releasing large quantities of methane into the atmosphere. The level of methane emissions depends on the quality of coal and the method of extraction. Coal buried under high pressure in the earth can retain more methane. Surface mining releases around 10% of the methane per unit emitted by deep mining.

The safe transport of precious energy resources is of utmost importance for public health and the environment. The fortuitous release of all fuels during transportation is extremely hazardous and strict measures must be taken for its prevention. Oil spills are particularly dangerous. Accidents occurring to oil tankers, which may release devastating hydrocarbons into the oceans, are disastrous and should absolutely be prevented.

During the combustion phase of fossil fuels, in addition to energy production, a high number of by-products are also generated, including carbon dioxide (CO_2), carbon monoxide (CO), methane (CH_4) and other hydrocarbons, nitrous oxide (N_2O) and various nitrogen oxides (NO_x), particulate matter and certain metals and radionuclides. The carbon content varies between the types of fossil fuels. Coal emits around 1.7 and 1.25 times as much carbon per unit of energy as natural gas and oil, respectively.

Uranium, the heaviest natural element, is radioactive and occurs naturally in low concentrations. The EU possesses hardly 2% of the global known reserves. Fissionable nuclear fuels are remarkable sources of energy. The energy released from the fission of one atom of uranium, is, for example, several million times greater than that produced by the combustion of one molecule of gasoline. The fission process utilises only a small part of nuclear fuel. Once separated from the waste, amounting to around 4%, recovered uranium and plutonium can both be used again to generate more electricity. Material obtained from the decommissioning of nuclear weapons can also be recycled as nuclear fuel. Thorium could also be used, in the long term, as fuel for nuclear fission reactors.

Renewable energy sources, including hydropower, solar, wind, biomass and biofuels, geothermal, ocean and tidal energy, are fundamental vectors towards a sustainable energy future and offer special opportunities for cities. They accounted for almost 7% of the EU energy supply in 2006. Countries that use renewable energy to a significant extent include Portugal, Finland, Austria and Sweden and cities are particularly interested in moving towards decentralised energy production based on renewable energy sources. Sweden and Austria produce more than half of their electricity from renewable sources. In Amsterdam, 37% of households use green electricity.

Primary energy sources are finally transformed into electricity, heat and motion. This transformation involves an array of critical losses, accounting for as much as a third of all the energy used. The final energy consumption, the energy actually used by consumers, is lower than the primary energy consumption and the difference is mainly accounted for by the huge losses in conversion and distribution.

Energy consumption was stabilised in the EU27 over the last few years, demonstrating a disassociation from GDP. In 2006, the overall energy consumption in the EU was covered by coal (16%), oil (40%), gas (22%), nuclear energy (15%) and renewable energy sources (7%). Although industrial energy consumption is being stabilised, mainly as a result of the transition to a digital, knowledge-based and service-oriented economy, demand for transport and electricity and heat from households and the tertiary sector is increasing.

Energy consumption patterns in countries and cities have key differences but also many common points. Everywhere, overall energy consumption growth is generally disassociating from the increase of GDP, much more than the use of other goods and the production of waste. Oil and oil products are the main source of energy in all EU countries except Sweden and France (nuclear energy) and the Netherlands (gas).

Gas is the second most commonly used fuel, although it represents a small percentage of the total energy mix in Sweden, Greece and Portugal. Renewables are the second largest source of energy in Finland, Portugal and Austria and the third largest in Sweden. Great differences among member States seem to occur not only in the composition of the energy mix, but also in the per capita consumption of energy. Luxembourg has the highest level of consumption, followed by Finland, Belgium and Sweden.

Energy consumption per capita reveals links between energy demand and the maturity of the economy. It is also linked to climatic and cultural features, population density and the structure of human settlements. Northern countries, with colder climates and low population densities, such as Sweden and Finland, are huge energy consumers. Countries with a warmer climate are installing a growing number of cooling systems and the seasonal distribution of energy consumption is changing. The severity of winters is also reflected in energy consumption peaks.

Industry accounted for 28% of total energy consumption in 2009. Investment in technology and innovation has enabled European industry to become more eco-efficient and advance towards the stabilisation of energy consumption. All energy-intensive industrial sectors demonstrated significant reductions in energy consumption during the last few years. Environmental regulation and enforcement have been key driving forces.

Energy consumption by the residential and services sectors grew at a moderate rate, as improvements in energy efficiency were partly offset by a systematic rise in levels of material comfort and an increase in the number of (smaller) households. The result has been higher per capita consumption, with energy used for space heating falling slightly and energy used for electricity rising. More than half of the total household electricity use for appliances and this part is growing (IEA 2003b).

Energy consumption by the transport sector, depending almost entirely upon oil (97% of transport consumption, representing 67% of final oil demand) has increased steeply. The transport sector accounted for more than 30% of total consumption in 2006 and it is the fastest growing energy consumer in the EU. This is mainly due to the continuing growth of road transport, passenger and freight. Air transport is also increasing dramatically, owing to the increase in the number of leisure trips. The shares of rail and inland waterway transport are declining and the modal split is shifting towards less unsustainable means, mainly road transport. Most price structures tend to favour private over public transport.

Unlike energy consumption, electricity consumption per capita in the EU is still increasing in line with GDP. In 2005, electricity in the EU was generated from nuclear energy (35%), solid fuels (26%), oil (8%), natural gas (16%), hydropower and other renewables (15%). Electricity production is predominantly centralised and characterised by dramatic losses. The electricity systems have been the bloodstream of industrial societies. "Business starts with power" has been a reality for a long time, but the classical electricity generation is now considered to be outdated and inefficient. The capacity to be installed over the next 20 years both in developing countries to facilitate access to modern services and in developed countries to replace old power stations and meet increasing demand holds many opportunities for radically improving electricity generation.

An essential change in electricity supply is the transition towards networks of smaller decentralised power plants nearer the consumers. Decentralised energy production can be much more efficient, as it allows the financial costs and energy losses associated with the long-distance national transmission systems to be radically reduced. Micro power generation is expected to continue developing gradually alongside the grids, as mobile telephony, and increasingly incorporate renewable energies (Greenpeace 2005).

Heating and, increasingly, cooling of buildings, account for about one third of total energy consumption. The needs range from office and household heating and cooling, including hot water, to steam production for industrial uses. Unlike electricity production, heat production is predominantly decentralised, whether it takes the form of individual heating systems or of dedicated heat stations with their associated networks.

Combined heat and electricity generation can lead to spectacular energy savings, as demonstrated in countries such as Denmark. A combined heat and power system enhances the role of waste energy from electricity production, while it helps to avoid the environmental impacts from additional heat generation. The overall system can reach very high efficiencies owing to the inherent characteristics of the process. To fully enhance the benefits, a sufficient heat load must be connected to the system. The produced heat can be used locally for district heating and industrial processes. Trigeneration, combined heat and power generation with additional production of cooling, promises further improvements.

Combined heat and electricity generation and decentralised district heating are not a novelty for European cities. Saarbrücken installed its first district heating system in 1964. It uses cogeneration plants for almost all its electricity production. More than 50% of electricity production in Denmark and the Netherlands are provided by combined heat and power systems.

Affordable prices are important for energy and are at the centre of much government and city attention, especially in times of crisis. Fuel poverty is usually referred to as the penultimate stage of poverty in which people avail of food and housing, but cannot purchase energy. In Middlesbrough, in the UK, 22% of households belong to this category, as evaluated with the aid of the research tool "fuel poverty indicator", developed by the University of Bristol. The project "Warmer Homes, Warmer Hearts" set up an affordable warmth strategy to help citizens to move out of fuel poverty and into affordable warmth.

The liberalisation of the energy market can lower electricity prices and bring more and better products and services, e.g. green electricity. It creates a challenge for new investment in cleaner technologies and energy efficiency. As new investors enter the market, such as medium-sized renewable schemes, suppliers are forced to cut costs. However, market liberalisation may make energy companies reluctant to invest in new, often more expensive and risky technologies, such as cogeneration. In Germany, liberalisation is deemed to have discouraged the shift to combined heat and power systems, which are capital-intensive.

Many European cities have joined their efforts to radically improve their energy performance. The EC-supported CONCERTO initiative brings together dozens of cities from across the EU to improve together their energy efficiency. Most efforts focus on the integration of the use of renewable energy sources with effective energy-efficiency measures. CONCERTO members work on areas as diverse as eco-buildings, distributed energy from renewable sources, energy-efficient building design and management, cogeneration of heat and power and clean district heating. Cities and observers share experiences on the internationally most advanced concepts and technologies and assess the potential of different policy measures. The final stage is the issue of policy recommendations, from the cross-evaluated experiences and exchanges.

Under the CONCERTO initiative, the Polycity project focused on energy optimisation and the use of renewable energies in three large urban areas in Germany, Spain and Italy. The German project is Scharnhauser Park, a former military area recently developed and hosting 6,000 people. The project is designed as an exemplary ecological community development, with low-energy buildings and a new biomass cogeneration plant. Low-energy building standards have been prescribed for all new constructions and connection to the heating network of the biomass cogeneration plant is obligatory. One of the aims is to use the residual heat to cool the office buildings in summer. Furthermore, all buildings will be equipped with optimal heat insulation to minimise heat loss in winter.

Another noteworthy pilot project is taking place in Cerdanyola del Vallès, a development area planned for 50,000 inhabitants north of Barcelona, and the Arquata in Turin, a labour district which is being restored according to ecological criteria. Each project is embedded in a network of regional partners and observer communities to guarantee the healthy development of the project and an effective exploitation of the results. The Polycity project is flanked by an eastern European and a western Canadian network of cities and communities, willing to take up and disseminate the experiences and results. Special training workshops are foreseen to provide the necessary information to communities wishing to replicate the projects.

4.2   Distributed Energy Systems and Renewable Energies

Microgeneration means that every home and district can become a mini power station and complement the grid. Power loads between neighbourhoods could be balanced and blackouts be avoided. The integration of decentralised energy resources and renewable energy into the main electrical grid is expected to change the energy paradigm of urban societies, with electricity generated in large power plants and delivered to the consumers through a passive distribution infrastructure. The benefits can be impressive and their optimisation requires the collaboration of all stakeholders, including utilities, independent power producers, central and local governments, regulators and the industry.

Distributed energy systems and renewable energy sources are key axes for a sustainable energy future. Renewable energy applications for electricity generation, heating and cooling and as biofuels for transport involve different processes and technologies maturing at different rates. Some have already penetrated the market, others are being deployment, whereas others are just demonstrating their potential (IEA 2008b).

Renewable energy sources are indigenous and contribute to the security of energy supply but are unevenly distributed throughout the EU. The obligation to increase the share of renewable energies in the energy mix is expected to create new market and business opportunities, especially for small and medium-sized enterprises, while endowing remote and island communities with assets and benefits. In 2006, renewable sources contributed 15% of electricity generation in the EU. Hydropower accounted for 87% of the total electricity produced from renewable energy sources. Biomass remained the second most important contributor to electricity from renewables and is particularly significant in Finland.

Renewable energies such as biomass, solar and geothermal energy have a huge potential in the heating and cooling sector, but they only account for 10% of total heating and cooling. There is an absolute need to integrate renewable technologies into the mainstream heating and cooling systems and widen the use of biomass-fired combined heat and power plants.

Establishing policies for renewable energy markets, expanding financing options and developing the required capacity were suggested as the main policy priorities at the World Renewable Energy Conference (Bonn, June 2004). The high-level conference asked for the radical promotion of renewable energy across the globe as a means to combat climate change, secure energy supply and reduce poverty. The ambitious 20% EU target for 2020 may help renewable energies to realise a greater part of their potential and increase their share of the market.

The EU is a world leader in renewable energy. Annual growth rates have been impressive, and costs have decreased and production has increased. The economic and social importance of the sector is significant. Renewable energy in Europe has a noteworthy turnover and provides many hundreds of thousands of jobs. However, the development has been uneven throughout the EU and given that the external costs of the fossil fuels are not taken into account in tariff structures, renewable energies are still not considered to be competitive in relation to conventional energy sources.

4.2.1 Wind Energy

The EU has excellent wind resources. Framework conditions enabled private sector competition and led to rapid technical progress, increasing performance and competitiveness. The developments in the EU countries, which encouraged a wind energy policy, demonstrate that publicly funded financial incentives, such as fiscal measures, concessionary pricing and capital grants, are easy to implement and can be very effective.

The Wind Day on each 15 June, organised for the first time in 2007 by the European Wind Energy Association, encourages the use of wind energy. During the years 1995–2005, cumulative wind capacity in the EU increased by an average of 32% per year. All scenarios of possible wind energy developments in the coming decades suggest continued rapid growth.

Onshore, very large wind power plants are being developed or are under discussion. The future of the wind energy industry depends on technological progress, institutional change and market dynamics, including the price of fossil fuels. For several countries where the market stimuli make wind power attractive, the main barrier is the difficulty of obtaining land-use planning consent. The role of local and planning authorities is very important, since a great number of projects fail at the stage of permit request. In the Netherlands, about two thirds of all initiatives fail to get building permits (EWEA 2001).

The next major phase is the exploration of potential wind energy offshore. The first floating turbines have already been realised. The best EU wind resources are offshore and are largely undeveloped. Offshore wind farms are very costly, but they cause near-zero scenic impact and this could allow the construction of large wind farms without friction with local communities. The greatest challenges relate to cost-effectiveness, design, reliability and power electronics. Their large scale could offset the cost disadvantage.

Projects, already developed in the UK, Denmark and Sweden, provide lessons and models. Examples include the DOWNVInD (Distant Offshore Windfarms with No Visual Impact in Deepwater) offshore wind farms incorporating a demonstrator project to install and monitor two wind turbine generators in deep water off the coast of northeast Scotland. The Dutch Amalia wind park, located approximately 15 km offshore from the coast, is one of the largest of its kind in the world. The offshore high-voltage station was successfully installed in early May 2007. The wind park comprises 60 turbines each with 2-MW capacity, generating approximately 400 GWh per year, which is enough to supply 125,000 households.

4.2.2   Solar Energy

The sun is the world primary source of energy. Solar energy has the highest theoretical potential for energy production. Solar energy systems can harness solar rays and produce electricity and heat. Technologies to capture abundant but diffuse solar energy are less mature than those for wind energy. Solar radiation reaches the earth with a density which is adequate for heating but not for electricity. Concentrating solar power technologies and systems include parabolic solar collectors, solar tower power plants and solar dish/engine systems. Technology and design are improving constantly and research is addressing issues related to efficiency, the intermittent character of the solar energy and storage.

Solar energy has a huge potential and represents a real chance for distributed energy in sunny countries and also in the developing world, where micropower is often cheaper than extending the grid. However, high costs still restrict solar energy to marginal and niche applications, such as telecommunications, leisure, lighting, signalling, water pumping and rural electrification (EUREC 2002).

Solar photovoltaics and thermal power generation still have high investment and operation costs. Photovoltaics use solar cells to convert light directly into electricity. The energy produced goes directly to the grid or is stored in batteries. Photovoltaics still represent an expensive solution even though they enjoy high reliability, a long lifetime, modularity and low maintenance costs. National programmes, such as the German 100,000 roofs, help stimulate the technology and gradually bring down the cost.

Photovoltaic electricity generation has a low energy density per surface area (around 1 GWh per hectare per year) and the efficiency does not depend on the size of the plant. Photovoltaics can be used in stand-alone, hybrid and grid-connected systems. The intermittent character of photovoltaic electricity generation, depending on solar radiation, is influencing their integration into the energy supply. In the case of connections to the grid or a minigrid, the latter may serve as a backup system. In stand-alone applications, a storage and conditioning system has to support the photovoltaic system, which can also be combined with small power systems as fuel cells.

The integration of photovoltaic systems into the built environment represents an attractive way to generate local renewable energy. Apart from producing energy, photovoltaic louvres in front of glass facades and windows can provide shade from direct sunlight and prevent the overheating of places while allowing the passage of sufficient daylight. The intelligent aesthetic and functional incorporation of photovoltaic modules in building roofs or facade structures can minimise visual intrusion and significantly reduce the cost of the systems.

For solar thermal power systems, the main barriers include physical constraints, due to the requirements for direct radiation, the high installation and production costs, performance and reliability.

In Spain, Seville has the first commercial concentrating solar power plant in Europe, the PS10, inaugurated in 2007. PS10 was designed to produce electricity for a population of 10,000 people and avoid the emission of about 16,000 tonnes of CO_2 per year. The plant has more than 600 movable mirrors concentrating solar radiation onto the top of a tower through a solar receiver and a steam turbine.

Barcelona proposed solar energy as a cardinal element in its 2002–2010 Plan for Energy Improvement. The plan includes demonstration and promotion projects and foresees the integration of the energy components in urban development. The Solar Thermal Ordnance of Barcelona has inspired many Spanish cities and had a major impact on the new Spanish building code. From its enforcement and until the end of 2006, more than 40,000 m^2 of solar panels had been installed, achieving annual savings enough to provide hot water for 60,000 inhabitants.

4.2.3   Bioenergy

Bioenergy sources of energy can include residential organic waste, agricultural and forest residues and algae. Biomass is versatile and can generate electricity, heat and transport fuel. It could, in addition, valorise waste and transform a liability into an asset. At the world level, biomass is the fourth largest energy source, able to provide a variety of energy products and services, including heat, electricity, gas through gasification and digestion of wastes and liquid fuels through fermentation and synthesis after gasification. However, most of the biomass resource is non-commercial and only a small percentage is used in modern processes, to produce electricity, steam and biofuels. In the EU, Austria, Finland and Sweden are the leaders in biomass utilisation (EC 2000).

The share of biofuels in the energy mix of the EU is still small, the principal obstacle being the price difference with fossil fuels. Biofuels can be primarily divided into biodiesels (70–80% extracted from organic oils and sunflower) and alcohols extracted from beetroot, wheat, etc. Among the production options, preference is being given to high-yield crops with low intermediate input and no effect on biodiversity. Biodiesel could be used without any major technical problems to replace diesel. Alcohols can be mixed with conventional petrol up to a level of around 15% without any technical modifications required for the vehicles (IEA 2004b).

Biogas sources include recovered gases from sewage, landfill sites and agricultural waste. Their use could lead to the reduction of methane emissions. The environmental impact of biomass energy is dependent on the primary source. The use of waste is the most beneficial. Although the use of wood or crops is still CO_2-neutral, as the other types of biofuels, the implications for land and fertilisation necessary for their continuous supply reduce their environmental benefits.

A major disadvantage for the development of the first-generation biofuels is their competition with food crops and with high-biodiversity ecosystems. In 2007, the dilemma of "fuelling cars with corn-based ethanol for 800 million drivers, versus feeding two billion people with corn" became prominent.

Following a year of rapidly rising food prices and global rioting over higher food prices, the EU reconsidered its biofuel production targets to include more non-food biofuel sources. The second-generation biofuels, made out of agricultural waste, offer a promising avenue for transforming a liability into an asset.

Amsterdam is already using biomass from municipal waste and sewage sludge to generate green electricity and heat. The metro and trams operate on this electricity. The Waste and Energy Company is already a large producer of sustainable energy. More than half of the incinerated waste, of non-fossil origin, is considered to be biomass (Amsterdam Climate Office 2008).

The EU target that fuels from renewable energy sources (biofuels, hydrogen, renewable electricity) should make up 10% of road transport fuel by 2020, providing they can be certified as sustainable, can give a decisive boost to the development of the second-generation biofuels and increase their share of the market.

4.2.4 Fuel Cells and Hydrogen

Fuel cells are electrochemical devices that convert the energy of a chemical reaction into electricity and heat. They produce energy from hydrogen and oxygen in a much more efficient and cleaner way than conventional combustion engines. Unlike batteries, they do not store energy, but support a continuous flow process. Sir William Grove constructed such a cell in 1839, with oxygen and hydrogen reacting on platinum electrodes to produce electricity. However, the development of fuel cells has only taken off in the last 30 years. The main breakthrough was the radical reduction of the size of the fuel cells needed to run a small car. It is expected that fuel cells will replace, in the medium and long term, a large part of the current combustion systems in industry, buildings and road transport. In the long term, fuel cells and hydrogen are expected to form an integral part of energy supply based on renewable energy systems. This may lead to a significant international market for fuel cells in transportation and stationary applications.

The fuel choice for fuel cells is an important issue. In the long term, hydrogen from renewable sources is the ideal fuel. In the medium term, fuels for fuel cells will have to be provided by fossil fuels such as natural gas, biomass and possibly coal. The drawback of CO_2 emissions related to fossil fuels can be avoided by CO_2 capture and underground storage. The energy generation process is intrinsically clean and efficient. The conversion produces only water and, if a fuel processor is required to reform a hydrocarbon fuel to hydrogen, only small amounts of CO and NO_x are produced. The efficiency is high, because the electrochemical conversion process is not limited by the same physical laws of thermodynamics that govern combustion processes.

In general, fuel cells are composed of an electrolyte, which performs the electrochemical reaction. Hydrogen gas is released at the anode and oxygen is released at the cathode, and the energy generated forces the released electrons into an external circuit. The resulting electrical energy is produced as a direct current and an inverter converts the unregulated output into a regulated alternating current supply. The main by-products of this process are heat, which can be reclaimed if a combined heat and power system is used, and water. To generate a substantial output, fuel cells are arranged in stacks with parallel electrical connections.

Fuel cells vary depending on the type of electrolytes used, for example alkaline or direct methanol fuel cells, and the stack's operating temperature. On the whole, fuel cells can be distinguished as belonging to low-temperature and high-temperature cells. Low-temperature cells operate at temperatures below 100°C and have a short start-up time, but have very specific fuel requirements. High-temperature cells have a long start-up time and often need heat insulation; they are, however, able to utilise an extensive range of fuels.

Hydrogen is a key energy carrier for a future sustainable energy economy. It is abundant and perfectly clean. It provides a unique pathway for gradually reducing today's dependency on fossil fuels and increasing the contribution of renewable energy sources. Used in fuel cells, it combines with oxygen electrochemically, without combustion, yielding an electric current, heat and water. A fuel cell generates electricity as long as hydrogen and oxygen are fed to it. Hydrogen can be used in fuel cells for all end-use applications, where it is highly efficient and intrinsically clean, and in gas turbines for large-scale electricity production.

Hydrogen can be produced from fossil fuels such as natural gas, by electrolysis of water with electricity from renewable energies such as hydropower, wind, solar energy and thermal energy and, in the long term, from photovoltaic electricity. In a long-term energy supply with a large share of renewable energy, hydrogen will play a key role in adapting the future energy supply, predominantly produced in the form of electricity, to energy demand (IEA 2003b).

An important objective is to reduce the cost of hydrogen production to make it competitive with the cost of conventional fuels. For the medium term, hydrogen from natural gas is the most likely route to achieve cost-effectiveness. The drawback of CO_2 emissions could be addressed by capture and underground storage.

Hydrogen can also be produced from water by electrolysis. For this, it is crucial to have cheap electricity from renewable sources, such as hydropower in Iceland, as the cost of electricity accounts for 80% of the cost of hydrogen production by electrolysis. Nuclear power is also being investigated as a possible source of hydrogen production, either as a provider of electricity for water electrolysis or as a supplier of high-temperature heat for the thermochemical decomposition of water.

Cost-effective transport, distribution and storage of hydrogen are major issues, together with the creation of an appropriate infrastructure. Regulation and standardisation of the hydrogen infrastructure on a European level are crucial. Hydrogen infrastructure, distribution, storage and utilisation are subjects of technological development. The potential for the storage of hydrogen in carbon nanofibres is deemed to provide fuel cell vehicles with a large autonomy.

Many cities bet on hydrogen for their public transport. Amsterdam wishes to acquire a position as one of the leaders of the hydrogen world. Three buses were already operating on hydrogen in 2008 and the first hydrogen boat has been inaugurated. In cooperation with other urban regions, it serves as a test site for large-scale transport operations (Amsterdam Climate Office 2008).

4.2.5   Geothermal, Ocean and Tidal Energies

Renewable energy sources also include geothermal, ocean and tidal energy. Geothermal energy flows from the hot interior of Earth to the surface, where it is lost by radiation into space. Global geothermal resources, using today's recovery and utilisation technology, contribute 1.6% to global electricity production, just after hydropower and biomass. In the EU, electricity generation from geothermal resources in Italy accounts for 1.7% of the total production.

In Europe, more than 100,000 homes are heated by geothermal energy. The EC-supported "European Hot Dry Rock" project, bringing together partners from France, Germany, Italy and Switzerland, tries to enhance widened natural fracture systems. It injects water at high pressure which is then heated and returned to Earth's surface through various production wells. A heat exchanger transfers energy to a second circuit that drives a turbine generator to produce electricity.

Oceans cover three quarters of the surface of the planet. Ocean, wave and tidal energies represent some of the most plentiful sources. However, ocean and tidal energy are expensive to extract and subject to storm damage. Tides have a large potential to provide energy, but the ocean is proving to be a difficult energy source to harness. Well-defined and selected wave and marine pilot projects and more detailed resource assessment could be decisive for the future of ocean energy.

Ocean energy industries and scientific communities are small and scattered. The European teams developing wave and tidal devices, which convert energy generated by the waves and tides, are world leaders. The EC-supported research project "Wave Dragon" in Denmark, is the world's first offshore wave energy converter producing electricity for the grid. Moored in water, the Wave Dragon recovers energy that is generated by overtopping waves. The water is initially stored in a reservoir and is then passed through turbines which produce electricity. The prototype is a quarter of the size of the full system and the new technology is competitive compared with traditional hydroelectric power stations.

4.2.6 Realising the Potential of Renewable Energy Sources

Renewable energy sources hold much potential for the security, diversification and sustainability of the overall energy supply. The realisation of this potential requires significant initial investment, as was the case for other energy sources, such as coal, oil and nuclear energy, which benefited from considerable state aid. Governments may utilise various measures to encourage the use of electricity from renewable energy sources (RES-E), including tax exemptions or reductions, direct price support, and investment aid. Direct price support schemes are widely used in member States, whereby renewable electricity producers receive financial support in terms of a subsidy per kilowatt-hour provided and sold.

The administrative and planning procedures to be respected by potential RES-E generators can be source of multiple obstacles. The EU directive, adopted in 2001, suggests that assistance to renewable electricity producers should be improved and that municipal administrations have a role to play. The existing procedures should be thoroughly investigated to identify any aspects that may reduce the regulatory barriers for RES-E generators. For example, a fast-track planning procedure or more specific planning guidelines and assistance could facilitate green power investments.

Cities have also to ensure that RES-E generators, particularly small-scale suppliers, are easily able to connect to the transmission grid. Renewable electricity generators, predominantly those in remote locations, may be required to use expensive installations to feed their electricity into the grid. This results in higher and even prohibitive investment costs.

National, regional and local regulations need to be adapted to give clear priority to the installation of generation plants for RES-E. The UK climate change policy includes a scheme for the promotion of renewable energy, through fundamental and industrial research and precompetitive development activities. The aim is to encourage the development of an international competitive renewable and sustainable energy industry.

The playground for the various energy options is notoriously uneven. The main barrier for renewable energy sources to penetrate the market has been the non-internalisation of external costs in energy prices. The EC-supported ExternE project, which was undertaken by researchers from all EU member States and the USA, was designed to quantify the socioenvironmental costs of electricity production. It tried to assess plausible financial figures against damage resulting from different forms of electricity production (fossil, nuclear and renewable) and offered a scientific framework for issuing the EU guidelines on state aid for environmental protection. The results highlight that if all costs were taken into consideration, coal should be

2–15 times more expensive, oil 3–11 more expensive and gas 1–3 times more expensive. At the other end of the spectrum, the cost of photovoltaics should be 0.6 times the current price and the cost of wind energy should be 0.05–0.25 times the current price.

The energy plans of cities make increasing use of renewable energies and decentralised systems. In London, the opportunities for decentralised energy supply range from solar panels on residential buildings to local combined heat and power systems supplying thousands of businesses and homes. According to the mayor, an energy revolution is needed to overhaul the regulatory and commercial frameworks in place, provide incentives for combined electricity and heat and reduce the bureaucracy and costs presently associated with the individual and local energy production (Greenpeace 2005).

Many cities may be interested to follow the example of Samsø, a little Danish island with 22 villages, which used to be known for its strawberries. It is now known for its wind of change, a socioenvironmental movement with spectacular results. Until the late 1990s, the island's 4,300 inhabitants had a conventional attitude towards energy. Most inhabitants used electricity imported from the mainland via cable, much of which was generated by burning coal, and heated their houses with oil, which was brought in on tankers. Then, the residents of the island decided to change this situation. They formed energy cooperatives and organised seminars on renewable energy sources. They removed their furnaces and replaced them with heat pumps. By 2001, fossil-fuel use on Samsø had been cut in half. By 2003, instead of importing electricity, the island was exporting it, and by 2005 it was producing from renewable sources more energy than it was using.

The Dundee Sun City campaign aims at making Dundee Scotland's first solar city exploiting solar energy to the maximum. In Aachen, already in the early 1990s, two feasibility studies demonstrated that the wind energy potential was sufficient to cover 10–12% of the city's needs, and efficient solar panels on all the south-facing roofs could supply 55% of these needs. Owners of wind or photovoltaic equipment receive a guaranteed payment per kilowatt-hour for 15–20 years. The overall cost was included in the cost calculations of the municipal power utility and passed on to all consumers (Énergie-Cités 2001).

Renewable energy is an absolute priority for Freiburg. The city initiated, in 2007, a conference series on the role of local governments for the promotion of the generation, supply and use of renewable energy in the urban environment. Freiburg has learned many lessons through its successes and failures since its first energy action plan issued in 1986. The latest edition, the Climate Protection Strategy 2030, aims to reduce CO_2 emissions by 40% by 2030. Developed in 2007, the strategy has a strong focus on energy conservation, efficiency and technology, including the cogeneration of electricity and heating and cooling. The city has built up an impressive range of expertise and implemented excellent and diverse examples in the field of sustainable energy.

Various EU schemes promote wider citizen involvement. The EC ManagEnergy initiative supports work on energy efficiency and renewable energies at local and regional levels through training workshops and online events. The EC Sustainable

Energy Europe campaign has been designed to raise public awareness about sustainable energy production and consumption and help citizens to take an active part in changing the energy landscape. The annual awards competition, an integral part of the campaign, provides an invaluable opportunity to reward the most inspiring and worthy projects. From the 1,000 photovoltaic roofs in Salerno to the Bristol annual reduction of emissions targets of 3% and its sustainable fund LEAF to finance energy efficiency improvements, many examples merit quoting (EC 2008c).

4.3   Cleaner and More Efficient Energy Consumption

The transition to cleaner and decentralised energy systems and inexhaustible energies demands cities and governments define long-term policy objectives and alter the relationship between the supply of energy services and environmental degradation. The market is, in general, willing to collaborate in the development of the most profitable cutting-edge environmental technologies. Cities need to support efforts, e.g., in demonstration projects, intellectual property regimes, competition policies, education and training, financial and fiscal policies to enhance the availability of capital to innovative firms and communication and consultation to improve transparency and accountability.

4.3.1   Producing More Cleanly and Smartly To Ensure the Transition

Producing smarter and consuming less and better energy are the main paths towards sustainable development. Technology is a major dynamic in all equations evolving around these directions. It can, first, contribute to the reduction of the level of use of energy services, such as heating, lighting and mobility. Technology is also critical in reducing the amount of energy required to generate a unit of energy services by increasing the efficiencies of both energy supply and energy end-use systems. The choice of fuel used for energy production is another area where technology can play a key role. Last but not least, technology can enable a decrease in pollutant emissions through their removal from combustion gases (EC 2002).

The traditional fuels will continue to dominate the landscape for some time, depending on the transition dynamics. Among the traditional fuels, the future of coal is questionable and its position in respect to its immediate competitors, oil and gas, is weak. It is solid and massive and has a lower calorific value than oil and gas and generates pollution at every stage of its exploitation. The management of its transport by sea is particularly difficult, even though it does not entail the same environmental hazards as the transport of oil and gas. The future of coal will largely depend on the development of techniques which make it cleaner and easier to use, such as gasification, and minimise its environmental impact, such as improved combustion technologies and CO_2 sequestration.

Oil is an essential component of energy supply responding primarily to an almost exclusive demand for traffic. Many oil fields are in decline or have reached the stage of needing remedial action to maintain production. Surveying technology and remote sensing have dramatically increased the amount of available oil, but new investments to access the exploitable oil are unlikely to be encouraged once the world enters cleaner pathways. Risk governance has to address many transition problems such as the decommissioning of ageing oil platforms and support structures on sea shelves.

The replacement of coal by natural gas has advanced in the EU. Investments in combined-cycle gas turbines are continuing. The extrapolation of past and present market trends highlights that an important part of electricity in 2020–2030 will be produced by natural gas. The pace at which gas resources will be exploited and depleted depends also on the price of oil and gas in global markets. Higher oil prices may induce more companies to invest in gas prospecting and production and exploitation techniques.

Conventional natural gas reserves could approach depletion in half a century. The exploitation of unconventional natural gas deposits, very large amounts of which are expected to be trapped in the deep seabed in the form of methane hydrates (clathrates), seems an attractive option. The transportation cost is the main barrier to the wider use of natural gas. Breakthroughs in pipeline technology, which can considerably lower the cost of a unit volume per kilometre transported, create new prospects.

Enhancing the potential of cleaner technologies is essential for cities. The traditional means of electricity generation already benefit from technological advances that allow increasing efficiency while reducing polluting emissions. Cogeneration and trigeneration promise further improvements.

In the transition to decarbonisation, technologies of CO_2 capture and sequestration for postcombustion capture and storage in oil- and gas-depleted fields and aquifers could result in important reductions in emissions. The underground storage capacity in the EU is sufficient to store the CO_2 emissions resulting from EU electricity production for 700 years. Available separation processes are best suited to large emitters, such as power stations, oil refineries and heavy industries.

4.3.2   Consuming Less and Better

Consuming less and better is an aim for buildings, accounting for more than 40% of total energy consumption in the EU, mainly for heating, lighting and equipment. Increasing living space per capita and higher levels of comfort and equipment for homes and offices have led to rising energy consumption. The impact of energy use in buildings is pervasive, since it is estimated that Europeans spend the major part of their time indoors. Studies have demonstrated that there is a large potential for cost-effective energy savings in this field, probably larger than in any other sector. As buildings in Europe have lifetimes of between 50 and 100 years or more, the largest potential for improving energy performance, in the short-term, is in the existing building stock.

The great majority of the housing stock of European cities dates from before the energy performance norms and is the origin of much wasted energy. In Amsterdam, only 8% of the housing units have been built since the introduction of the norms and almost half of the 385,000 housing units were built before the Second World War. In 2006, 34% of the total CO_2 emissions in Amsterdam were caused by household use of electricity, gas and heat, despite a decline of more than 1% per year in the average household gas consumption. This reduction, due to the efforts of the housing associations investing for many years in better insulation and high-efficiency boilers, and the much better energy performance of newly built houses, has been counteracted by an increase in the number of housing units and especially by an increase in electricity consumption of 2% annually.

Major stakeholders include home owners, private landlords (22%) and residents (27%), who make decisions about insulation, double glazing, efficient heating installations, good ventilation or the use of renewable energy. Housing associations, owning more than 50% of the housing units, are also important players along with tenant organisations. If all housing associations undertook serious renovations until 2018, a CO_2 reduction of 31% would be possible and this could reach 37% if these efforts were continuing until 2025. The reduction will be partly achieved by selling units, tearing down old housing and building new housing. Since February 2007, the municipality has headed the leaders' alliance in which the housing associations are challenged to achieve the best performance and participate in concrete initiatives such as the Step2Save project and the network for energy savings, an initiative of the Amsterdam Federation of Housing Associations. The municipality and the housing associations are jointly creating three model houses to reduce emissions (Amsterdam Climate Office 2008).

Ecological homes, districts and cities are not luxuries, but necessities, critical to the survival of the biosphere. Public buildings and privately owned buildings that are used by the public can act as pioneers and serve as models for intelligent resource-saving buildings. Municipal service buildings, hospitals and schools will become fields of experimentation with optimal use of climatic conditions, e.g. the most favourable indoor temperatures, and sustainable management of resources and energy. Symbols are important, such as the buildings hosting the Danish and Dutch ministries of the environment. The "Display" initiative of Énergie-Cités encourages the demonstration of the energy conditions in public buildings for awareness building.

Toronto's Better Buildings Partnership (BBP) is an innovative programme administered by the Energy Efficiency Office, which provides resources for energy-efficient retrofits and construction of new buildings. The programme includes three sector-targeted segments, one applying to multifamily buildings with more than six dwelling units, one for all municipal, academic, social and health-care buildings and one for new constructions and commercial developments. The BBP's primary goal is to reduce CO_2 emissions which come from the energy used primarily to heat, light and cool buildings. The benefits include lower utility bills, increased resale value, reduced greenhouse gas emissions and potentially better indoor air quality, leading to better health and productivity. Additionally, BBP helps building owners in the institutional and not-for-profit buildings sectors take advantage of Toronto's Sustainable Energy Fund, which offers zero-interest loans for energy conservation or renewable energy projects.

Sustainability ethics are increasingly reflected in new building regulations. In the EU, the Energy Performance of Buildings Directive (EPBD), issued in 2002, suggested a common framework of harmonised measures for the development of integrated energy performance standards, to be applied to new and existing buildings when renovated. The approach to energy performance standards covers efficiency aspects ranging from clean and efficient energy generation to insulation and installed equipment. An integrated method addressing all energy aspects facilitates the most effective and efficient combination of measures and offers a basis for designers and builders to be able to recognise and promote high standards.

The EC and European Parliament are considering recasting the EPBD to introduce a number of key changes. Certificates could be displayed in buildings larger than 250 m² that are occupied by a public authority and in commercial buildings larger than 250 m² that are frequently visited by the public. The energy performance of existing buildings of any size (present threshold of 1,000 m²) that undergo major renovations has to be upgraded to meet minimum energy performance requirements. The EC may establish common principles for the definition of low- and zero-carbon buildings. Requirements may include targets for an increase in low- and zero-carbon buildings with separate targets for new and refurbished buildings, new and refurbished commercial buildings and buildings occupied by public authorities.

The key recommendations offered from the CONCERTO network of cities highlight the role of the public sector as a front-runner in standard raising by displaying the certificate in a prominent place visible to the public and insist on including final energy use figures as well as the corresponding primary energy use and CO_2 emissions ratings in the energy performance certificate. The network of cities recommended lowering the threshold of 1,000 m² for existing buildings that undergo major renovation, to be accompanied by systematic technical, legal and financial supporting mechanisms. Finally, the network suggested defining a certified professional category, legally recognised for certification, to ensure the quality of inspections and energy performance certificates.

The EU portal BUILD UP tries to help building professionals, local authorities and citizens willing to share their experience on and improve their insight into how to cut energy consumption in buildings. The portal builds on best practices, tools and technologies available across Europe for an effective implementation of energy-saving measures in buildings.

Public lighting has often been a subject of controversy. Ample lighting has a beneficial effect on public safety and the atmosphere of a city. It has proved to help reduce both road accidents and road incidents. But it also results in high energy budgets. Intelligent light measuring and more efficient lamps and materials could help reduce energy use. Responsible public lighting would benefit from lighting adapted to the degree of darkness. The municipality of Amsterdam replaced older lamps with newer types and is switching entirely to electronic components. In 2007, the entire City Hall of Amsterdam was equipped with energy-efficient lamps, achieving 45% energy savings. The project earned the 2008 European Light Award. Furthermore, new-generation LED streetlights were tested and can provide energy savings of up to 51% (Amsterdam Climate Office 2008).

4.4 Energy Ethics: For Science with a Conscience

The overarching concept of sustainable development shed light on the obligation to preserve global capital and equity as regards future generations. Ethical issues in energy arise with the freedom of choice when faced with alternative courses of action. They are intrinsically linked to the right to energy for all and the options offered to citizens. The provision of energy services should not compromise the environment and the rights of others. Each energy option has incontrovertible weaknesses and strengths and presents simultaneously irreducible advantages and disadvantages. Government energy choices are conditioned by previous decisions with long-term effects, high-scale invested capital and long-lived infrastructures. Citizen choices are constrained by the options offered to them in precise places and contexts.

Sustainability ethics ask for energy strategies to facilitate the transition to a sustainable energy supply, through energy savings and the enhancement of renewable resources. The key reference on ethics in the European Union is the Charter of Fundamental Rights, adopted in 2000. It is an integral part of the Lisbon Treaty and includes provisions on dignity, freedoms, equality, solidarity, citizens' rights and justice.

The World Commission on the Ethics of Scientific Knowledge and Technology (COMEST) created by UNESCO has already published a ground-breaking report on the major ethical issues linked to access to and use of energy resources, including fossil fuels, nuclear energy and renewable energy sources. The report insisted on the importance of research in this field with the multifold objective to increase efficiency, effectiveness and equity and reduce the inherent risks. It also highlighted that the ethics of energy can only be comprehended in the context of a better understanding of how energy is woven into the social and economic fabric (UNESCO 2001).

Ethical questions may relate to the radical character and speed of technological change. The deployment of technologies for increased environmental health and safety, the fostering of local capacity to monitor developments and voluntary energy and environmental audits and agreements have important ethical dimensions. Reversibility is often pointed out as a main issue and attention is given to the distinction between the reversible or irreversible consequences of energy options in the short, medium and long term.

Specific energy issues, such as the retrievability of nuclear waste, have important transgenerational ethical implications. The precautionary principle also asks for the prevention of unexpected and unintended consequences that introduce new uncertainties and may have irreparable effects. The search for an optimal and consensual solution to the management of nuclear waste, taking into account potential long-term risks and the issue of definitive waste disposal versus irretrievability allowing future generations to take their own decisions, is subject to much debate.

Public perception of the energy options in the EU is being detected through opinion polls included in diverse Eurobarometer surveys, based on a questionnaire completed by a representative sample of the EU population. EU citizens asked to prioritise the main criteria for choosing energy options pinpointed the lowest risk of pollution for the future, highest price stability and greatest reliability in terms of supplies. According to these criteria, most EU surveys of the last decade suggest that Europeans most favour renewable energy sources. Nuclear energy comes, in general, in an intermediate position, whereas solid fuels and oil come last.

The Eurobarometer survey carried out in EU25 in 2006 highlights that, when considering the whole range of socioeconomic issues, energy is perceived rather as a "back of the mind issue". In fact, EU citizens rate energy issues (14%) far below unemployment (64%), crime (36%) and health-care systems (30%) as their main concerns. Concerning energy, protection of the environment and affordable prices for consumers were suggested as the top priorities by EU citizens. The survey reveals that, looking 30 years ahead, Europeans anticipate a fundamental swing towards the use of renewable energies. European citizens rank nuclear energy as likely to be the third "most used" energy source in 30 years' time, after solar energy (49%) and wind energy (40%), especially because of its small contribution to greenhouse gas emissions (EC 2006c).

For citizens, energy is most often associated with high prices. A third (33%) of Europeans spontaneously relate energy issues to soaring prices and 45% consider that their government should guarantee low energy prices as a top priority in their energy policy. The majority of EU citizens (61%) also think that their country is significantly dependent upon imported energy. The perceived degree of dependency falls to 53% when respondents are asked to consider overall EU dependency levels.

Concerning the energy options and citizen preferences for the future, the most frequently quoted criterion was protection of the environment and public health (72%), followed by low prices for consumers (62%) and uninterrupted energy supplies (30%). There is public enthusiasm for renewable energy sources, considered to be the most advantageous and environmentally friendly and, to a lesser extent, the most efficient source in 50 years' time. Support for renewable energy sources and nuclear fusion tends to depend on the level of education.

The majority of Europeans (59%) are not prepared to pay more for energy from renewable sources. However, 24% of respondents are still prepared to do so provided that the price increase is limited to 5%. There are significant differences between the EU15 and the 12 newer member States, with the latter being clearly more reluctant to pay higher prices for green energy. Five out of ten Europeans appear to be willing to reduce their energy consumption and 4% would make this change even if it implies paying more. Nevertheless, 16% of respondents were not willing either to change their consumption patterns or to pay more. When asked what the public authorities should do to help citizens reduce their energy consumption, a relative majority thinks that governments should promote the efficient use of energy more actively through information provision and tax incentives (EC 2006c).

EU citizens seem aware of the fact that nuclear power is one of the main energy sources in many European countries. Nuclear power ranks third among the most quoted used energy sources, after oil (81%) and coal (77%). However, their views are not completely accurate. In countries where nuclear power is the main source of energy, such as France and Lithuania, it is still only the third most chosen answer (78% in France and 49% in Lithuania).

The majority of EU citizens (58%) have heard of nuclear fusion, mainly in Sweden (99%), the Netherlands (86%), Denmark (72%), Germany (71%), France and Finland (69%). The analysis of the attitudes to producing energy from nuclear fusion clearly reveals that the issue is still difficult for the public to grasp.

The main sources of public concern in relation to nuclear energy include nuclear waste, safety and proliferation. The radical decrease of risks (probability × consequences) associated with nuclear energy has to convincingly address societal concerns and feed a thorough public debate. The reduction of real risks is a necessary but not sufficient condition for the reduction of the perceived risks. Risk assessment and management have a prominent role.

Safety issues are of great concern to Europeans, especially in relation to nuclear power stations, followed by food safety, safety at work and the safety of industrial sites. As far as energy-related research is concerned, EU citizens expect a significant impact on environmental protection and ask for more action with regard to renewable energy sources and cleaner means of transport.

Public opinion polls on nuclear waste highlight that EU citizens are rather poorly informed about radioactive waste. A 2008 Eurobarometer survey examined the attitudes of European citizens to radioactive waste and in relation to the results from three previous surveys conducted in 1998, 2001 and 2005. Radioactive waste remains a major issue for the acceptance of nuclear energy. Four out of ten of those opposed to nuclear energy would change their mind if there were a safe and permanent solution for the management of radioactive waste. Almost nine out of ten respondents would like the EU to monitor the situation and ensure a level playing field between member States (EC 2008d).

According to the 2008 survey, support for nuclear energy is highest in countries with existing operational nuclear power plants and in those where the citizens feel that they are well informed about radioactive waste issues. Nearly two thirds of European citizens recognised the beneficial effects of nuclear energy, such as diversification of energy sources, decrease in the dependence on oil and reduction of greenhouse gas emissions.

There is an overwhelming apprehension about the transport of waste to a deposit site, should this be built nearby, and the great majority of Europeans declare that they would be worried. They also confirm that they would be concerned about the harmful impact on their health and on the local environment caused by the potential building of an underground site near their place of residence. Nine out of ten

Europeans stated that they would be worried about the long-term safety implications of the deposit sites.

Governance has to address issues related to public resistance to certain energy options, which can raise difficult ethical considerations. Trust implications may result in certain technologies being rejected or inadequately developed. Increasing knowledge and public understanding of the social costs and benefits of alternative technologies involve agreeing on approaches for opportunity and risk management. Carefully targeted research in areas that offer concrete prospects for achieving a turnaround in public opinion may be most beneficial. The involvement of cities in setting policy agendas can contribute to developing technologies that respond to the needs, values and preferences of society. Restoring trust and respect is essential if policies are to be implemented effectively. Credibility is a key to acceptability.

Scientists top the credibility league as a source of information and insight. The majority of EU citizens consider scientists and environmental organisations to be the most trustworthy sources of accurate and useful information on energy issues. Energy companies rank fifth in the list, ahead of journalists and national governments.

The EU-supported ETHOS project, in the aftermath of the Chernobyl accident, in cooperation with the ministries for Chernobyl in Belarus, Russia and Ukraine, is exemplary. The enhancement of the living conditions in the contaminated settlements was one of the key objectives of the project. The main problems included the local dependency culture and the highly centralised character of remedial actions. A decentralised approach was developed and implemented, with the active involvement of local communities. The integration of human and social factors in the planning and implementation of countermeasures has proved to be decisive. The project demonstrated tangible improvements in socioeconomic conditions and contributed to the development of a coherent approach to effective contingency planning.

Risk governance, involving participation in assessing and managing hazardous activities, is receiving renewed attention. TRUSTNET, an EC-supported initiative, analysed the factors which influence the credibility, legitimacy and effectiveness of the scientific and regulatory framework and developed more coherent, integrated and equitable approaches to risk management. TRUSTNET advocated the "mutual trust" versus the "top-down and bottom-up" paradigm. Through a broad range of case studies and brainstorming seminars, it offered insight into the social management of risk. Sensitive safety problems, including tanker catastrophes and oil slicks, have not only to be addressed effectively, but also have to increase the feeling of security and promote citizenship.

A TRUSTNET case study focused on the involvement of stakeholders in the decisions concerning the energy future of Bavaria. The finance ministry organised a round table with stakeholders, representing all those potentially affected by the related decisions, to examine the energy options of the future and address issues such as market liberalisation, the phasing out of nuclear energy, decided by the German Government, and global climate objectives. The round table brought together 33 representatives from government, parliament, industry, energy producers and suppliers, consumers, environmentalists, trade unions and civil society for 1 year.

The move towards a sustainable energy path, supporting Germany's ambitious climate protection policy and the increase of energy efficiency were agreed as the main priorities. The future of nuclear energy provided the dividing line. Thirteen of the stakeholders were in favour of the extension of nuclear energy, especially to meet the climate-related objectives, whereas six suggested that abandoning nuclear energy is irreversible. A compromise was achieved in the form of postponing the decision until 2010.

The Chernobyl disaster underlined that a major nuclear accident cannot be ruled out, even though the safety requirements imposed on nuclear technologies in the EU offer a high level of reliability. The safety of installations involves establishing common standards for the remaining life of nuclear plants, improvement of inspection and surveillance methods and strategies for the prevention and mitigation of accidents.

By 2025, more than 60 of the EU nuclear reactors are due to be decommissioned. Decommissioning a nuclear power plant is a long process involving a cooldown period and the obligation to return the site to an environmentally acceptable situation according to the highest safety standards. The process may take up to 60 years and cost 10–15% of the original cost. The management and storage of waste and the exploration of innovative concepts offering advantages in terms of safety, cost and durability are essential.

Nuclear reactors in operation around the world are generally considered second- or third-generation systems, the first-generation systems having been retired some time ago. Experts suggest that the future of nuclear fission lies in the Generation IV reactors, currently the subject of intensive collaborative international research. The primary goals are to improve nuclear safety, expand proliferation resistance, minimise the use of natural resources and waste and decrease the cost to build, operate and decommission plants. Most of these designs are generally not expected to be available for commercial construction before 2030, with the exception of a version of the Very High Temperature Reactor called the Next Generation Nuclear Plant expected to be completed by 2021.

4.5 Prospects for Nuclear Fusion: A Competitor for Renewables?

Controlled thermonuclear fusion holds enormous potential, since it leads to a virtually unlimited source of cleaner and safer energy. It constitutes a frontier technology founded on a strong vision which requires substantial and sustained effort, not only in research and development, but also in international cooperation, socioeconomic modelling of options and financing. Fusion research spans various disciplines and relies on multiple innovations, achieved by cutting-edge developments in many scientific and engineering fields. Progress is being achieved by successive long steps that gradually lead into the next stride.

European research and development activities on fusion include studies and evaluation of alternative concepts of magnetic confinement, and coordinated activities in fusion technology, in particular research on fusion materials. Socioeconomic aspects of fusion, especially economic costs and social acceptability, are under continuous research together with safety and environmental aspects.

Nuclear fusion occurs naturally in stars. Although significant progress has been achieved in reproducing this concept on Earth with various fusion experiments, it was clear from an early stage that a larger and more powerful device would be needed to create the conditions expected in a fusion reactor and to demonstrate its scientific and technical feasibility. Scientists and engineers began developing conceptual and engineering designs for such a "next-step" device, at European and international level.

The international experimental fusion reactor ITER is a technoscientific megaproject that aims to advance nuclear fusion for the large-scale carbon-free production of baseload power. *Iter* also means "journey", "direction" or "way" in Latin, reflecting ITER's potential role in harnessing nuclear fusion as a peaceful power source. During its operational lifetime, ITER will test key technologies necessary for the next phase, involving the demonstration fusion power plant that will capture fusion energy for commercial use.

ITER is designed to produce approximately 500 MW of fusion power sustained for up to 1,000 s by the fusion of about 0.5 g of a deuterium/tritium mixture in its approximately 840-m^3 reactor chamber. ITER should generate the 500 MW of fusion power over periods of around 8 min, with a tenfold energy output-to-input ratio under conditions similar to those expected in an electricity-generating fusion power plant.

Joining efforts towards the international experimental reactor ITER has been a major step forward. It began in 1985 with a partnership bringing together the EU, the then Soviet Union, the USA and Japan. The Agreement on the Engineering Design activities of ITER, signed in 1992 under the auspices of the IAEA, between the EU, Japan, Russia and the USA, joined later by the other partners, has been instrumental. The conceptual and engineering design phases led to an acceptable detailed design in 2001, underpinned by research and development by the ITER parties to establish its practical feasibility. The USA opted out of the project between 1999 and 2003, and new parties included Canada, China, Kazakhstan, Korea and India.

The timescales are long and depend on the magnitude of investments. Initial studies indicated that slightly more than 8 years would be necessary between the start of ground excavation and production of the first plasma. The construction of ITER could last about 10 years. It would be followed by an exploitation phase lasting about 20 years and a deactivation phase lasting about 5 years.

Following international negotiations on the legal and institutional conditions of the establishment of an ITER legal entity, the EU conducted negotiations for the joint implementation, including construction, operation and decommissioning. The European site of Cadarache was selected as the site for ITER and the project moved on to its construction phase.

The commercial feasibility of fusion energy has to be ensured by market penetration. Models indicate costs which should make fusion competitive. Although nuclear fusion offers not to expose the public to any major risk of accidents and not to cause little long-lived radioactive waste, ethical issues are given great importance. Cities could play a major role for the social acceptability of nuclear fusion, a prerequisite for the exploitation of fusion energy, promoted as the energy of the stars on Earth.

In the second half of the twenty-first century, when the era of fossil fuels will definitely belong to the past, nuclear fusion, the energy of the stars, may be the only competitor to renewable energies in cities already in a much more advanced stage on their sustainability arrow.

Watercolour 5
Stockholm: Green Capital of Europe 2010

Stockholm: Green capital of Europe 2010

Chapter 5
Competitive and Attractive Cities

Make no little plans. They have no magic to stir men's blood...

Daniel Burnham (quoted after his death)

The future of cities depends largely on their capacity to generate and distribute wealth and the quality of life they offer to citizens. Cities are propellers of the economy, both the creators and the mediators of the dynamics that are released by the mutual reinforcement of activities clustering together. In times of crisis, the search for sustainable growth, in balance with the other components of sustainability, is critical. Multiwin innovations are essential for creating new assets, especially out of liabilities.

This chapter examines the competitiveness of cities and the assets that have to be enhanced for sustainable development. It argues that smart and green entrepreneurship and employment creation is crucial for going up the urban value chain. Innovative partnerships with eco-businesses can help reconcile green growth benefits with long-term sustainability goals. Quality of life and access to resources and knowledge are key factors of competitiveness, as highlighted by international surveys, assessments and rankings. The chapter ends at the competitive edge of the cities, their lungs and brains, the green and grey parks.

5.1 Cities, Players in the Knowledge Economy in Crisis

Cities are engines and magnets of the global economy. They constitute theatres of politics, networks of local networks and at the same time the nodes of global networks. They generate new ideas, technologies and services, manage a growing share of national and international exchanges and attract and direct both human and financial flows.

Cities are genuine protagonists on the scene of national and global economic operations. Economists regard them as both the creators and the mediators of the dynamics that are released by agglomeration economics and positive externalities, the mutual reinforcement of activities which cluster together (O'Sullivan 1996). Cities are the main generators of the wealth of the nations which enhance their position in the international sphere (Jacobs 1985).

V.P. Mega, *Sustainable Cities for the Third Millennium: The Odyssey of Urban Excellence*, DOI 10.1007/978-1-4419-6037-5_5,
© Springer Science+Business Media, LLC 2010

In 2005, the top 30 cities ranked by GDP generated around 16% of the world output. Tokyo and New York have estimated GDPs, in purchasing power parity, comparable to those of Canada and Spain, respectively (World Bank 2009). The EC Urban Audit revealed that for cities with more than one million inhabitants, the GDP is 25% higher than in the EU as a whole and 40% higher than the national average (EC 2007f).

Traditionally considered as crossroads of transactions and strongholds of production, cities became command and control centres, the places where economic flows can be decoded, condensed, intensified and finally converted into wealth. The service sector is by far the most developed as a source of income and employment in most cities. In the five largest urban agglomerations in the EU (London, Paris, Berlin, Madrid and Rome), more than 80% of all jobs are in the service sector (EC 2007f).

Globalisation brought extraordinary changes and altered inter- and intracity relationships. Information and communication technologies and the galaxy of the Internet developed to a point that do not simply affect and integrate economies, but impact on society, culture and lifestyles (Castells 2001). A global economy gives many more cities the opportunity to become assertive actors of a global city, but this world conglomeration might have strong central quarters and weak peripheral ones. Globalisation leads to the modification of the criteria for the adherence to a city and may affect the capacity of cities to respond to citizens' needs (Sassen 1994).

The competitiveness of a city depends on a multitude of factors, including the macroeconomic and institutional environment, microeconomic performance, openness to trade and investment, flexibility of the labour market, adequacy of physical and digital infrastructure, level of education and training, and ability to create and innovate. Higher competitiveness may result from increased productivity, better quality products or greater involvement and stimulation of a well-trained workforce. The information society provides new horizons and opportunities to be exploited. Many cities recognise that increasing competitiveness cannot come from compromises on social measures and environmental practices. Education is probably the most important single factor for sustained high performance.

Strengthening the management of urban infrastructure, reinforcing institutional capacity, improving the financial and technical capacity of municipal institutions and improving the regulatory frameworks are essential preconditions for better urban productivity. Governance is of key importance if benefits and profits have to be equitably shared among citizens. Improving productivity depends on the successful balancing of the various elements of macroeconomic policy managed at the national and the city level. It also requires leadership, openness and flexibility (ACDHRD 1995).

The knowledge economy and the conversion process from the industrial era have brought major challenges for most cities. Since the Industrial Revolution, urban landscapes have been endowed with manufacturing infrastructure and transport facilities linking them to places of consumption. Industries had to be restructured and relocated to generate productivity gains and meet the demands of international competition. Major world trends include the constant decrease of the contribution of manufacturing to the gross world product and the creation of clean, knowledge-based businesses and integrated services which flourish in urban environments.

As cities are subject to changes caused by the transformation towards an information society, they represent ideal places for actions that reinforce democracy through the use of new technologies. The Global Cities Dialogue, a worldwide network of cities founded in 1999, promotes an information society free of digital divide and founded on sustainable development. Mayors and high political representatives signed the Declaration of Helsinki "Mayors of the World for a Global Cities Dialogue on the Information Society" and committed to creating equal opportunities and access for all citizens.

National debate on urban and metropolitan issues is high on the agenda of the US administration which came into office in 2009. The creation of the White House Office of Urban Policy aimed at strengthening metropolitan areas as dynamic engines of sustainable prosperity. Problems were not only due to insufficient coordination but also to inadequacy of federal programmes, which inadvertently undermine cities and regions by encouraging insufficient and unsustainable development patterns. The US Conference of Mayors, representing more than 1,200 medium-sized to large city mayors, on the occasion of President Obama's first 100 days in office, praised the efforts of the administration to make US urban centres the springboard for the nation's economic recovery.

US metropolitan areas host 85% of jobs and 90% of economic growth. In closing the round table on urban and metropolitan policy, in July 2009, the American president recognised the special challenges facing cities during the recession. Four in five US cities had to cut services. He suggested that the US recovery plan has helped the economy and cities to weather the storm. But new bolder strategies have to be conceived and implemented for cities to achieve competitive, sustainable and inclusive growth. The first review in 30 years of the federal policy and funds to urban and metropolitan areas is being enriched with field visits to innovative cities try to identify projects that produce valuable and exemplary results (Obama 2009).

In October 2008, the American journal *Foreign Policy* noted that "as the world's biggest, most interconnected cities help set global agendas, weather transnational dangers, and serve as the hubs of global integration, they are the engines of growth for their countries and the gateways to the resources of their regions". In conjunction with the consulting firm A.T. Kearney and the Chicago Council on Global Affairs, *Foreign Policy* published a ranking of global cities, based on consultation with urban experts and thinkers. The rankings are based on the evaluation of 24 metrics in five areas: business activity, human capital, information exchange, cultural experience, and political engagement. The top ten cities were New York, London, Paris, Tokyo, Hong Kong, Los Angeles, Singapore, Chicago, Seoul and Toronto.

The first Japanese index of world cities was presented by the Institute for Urban Strategies at The Mori Memorial Foundation in Tokyo, after a comprehensive study of 35 global cities in October 2009. This ranking is based on 69 indicators covering the economy, research and development, cultural interaction, liveability, ecology and the natural environment, and accessibility. This Japanese ranking also breaks down the top ten world cities ranked in subjective categories by managers, researchers, artists, visitors and residents. It is interesting to note that New York, London, Paris and Tokyo are again at the top of the list, while Singapore, Berlin, Vienna, Amsterdam,

Zurich and Hong Kong occupy the next six places. According to the study, one implication of the growth of intercity networks is the emergence of an urban geopolitics. Urban leadership – at the political, corporate and economic level – is under more immediate pressure to act than national governments.

The EU Urban Audit highlighted that cities are the indisputable engines of economic growth across Europe. In virtually all European countries, urban areas are the foremost producers of knowledge and innovation, the hubs of the globalising world economy. Unemployment rates tend to be higher in cities. Across Europe, the unemployment rate was higher than the national rate in two out of three Urban Audit cities. Within cities, very large differences in unemployment rates can be observed between neighbourhoods (EC 2007f).

Like American cities, European cities have also been affected by the crisis that started in 2008 (CEMR 2009). The comprehensive plan to lead the recovery of the EU out of the crisis which started in 2008 has important actions of interest to cities. It includes short-term measures to boost demand, save jobs and help restore confidence and longer-term smart investments to yield higher growth and sustainable prosperity. Absence of action brings a risk of a vicious recessionary cycle of falling purchasing power and tax revenues, rising unemployment and ever-wider budget deficits.

The recovery plan puts forward concrete steps to promote entrepreneurship, research and innovation, including in the car and construction industries. It aims to boost efforts to tackle climate change while creating jobs through, for example, strategic investment in energy-efficient buildings and technologies. The targeted sectors account for a quarter of all jobs in the EU and are earmarked for priority action to ensure a greener future.

Public–private partnerships are key mechanisms for the implementation of the recovery plan. They all have special significance for urban economies and are expected to boost clean technologies through support for innovation. Public–private partnerships include three major initiatives, one on European green cars with combined funding, one on European energy-efficient buildings and one on factories of the future (EC 2009f).

Special emphasis is being given to knowledge-intensive green investments. Investing more in education and (re-)training for environmental skills may help people to obtain high-quality jobs and reintegrate the labour market. Investing in smart urban infrastructure and energy efficiency supports the construction sector, saves energy and improves efficiency. Investing in clean cars helps protect the planet and may create a leading edge in a highly competitive market.

The European plan builds on the Small Business Act and provides further help for small and medium-sized enterprises (SMEs), including removing the requirement on microenterprises to prepare annual accounts, easing access to public procurement and ensuring that public authorities pay invoices within 1 month. The plan also includes further initiatives to apply state aid rules in a way that achieves maximum flexibility for tackling the crisis while maintaining a level playing field.

The construction sector, the sector which, to a large degree, shapes and transforms cities, is recognised as a good barometer of the vitality of the economy, as it feels first the cold wind of a recession and the warm breeze of a recovery. The Architects' Council of Europe (ACE) initiated, in 2009, a series of quarterly surveys on the opinion of practising architects on the impact of the economic crisis on their offices and workload. The second survey was carried out during the second half of June 2009. Sixty-two percent of respondents said that the situation is "bad" or very bad" as compared with 46% in April 2009. The results of the survey reflect a relative pessimism; just 8% of respondents in June 2009 said that they thought the situation was "good" or "very good" as compared with 15% in April 2009 (ACE 2009).

Concerning the employment situation, nearly one out of three architects' offices had experienced a decrease in staff since the nominal start of the crisis in September 2008. The expectation of those who responded in June 2009 is that staff would decrease in 23% of offices in the coming 3 months. On the positive side, 7% of architects' offices had seen an increase in staff numbers since September 2008. The ACE asked for immediate increased public investment in sustainable construction and, in particular, the improvement of energy efficiency of existing buildings, which is in line with the concerted response provided by the European recovery plan (ACE 2009).

5.2 Aligning Green Growth with Sustainability

The role of enterprises in the transition of cities to sustainable development is essential. In search of sustainability, cities and enterprises have to become more open, innovative, flexible and inventive and make the optimal use of resources and technologies. Cities have to create the conditions for the optimal green products and services to come to the market. Enterprises have to deploy their inventiveness and make available environmental technologies, products and services with reasonable profits.

Cities traditionally have a much longer lifetime than enterprises. Antagonistic logics, with long-term political goals versus short-term economic profits, have long opposed cities and enterprises. Reconciling political goals with economic growth and profits is at the heart of strong sustainability. As entrepreneurship creates value moving from low-yield resources to higher-yield resources, urban sustainability also implies the constant raising of the scale of values. Large enterprises may lead to the edge city and SMEs have high potential for innovation and for revitalising urban areas. Cities can act as enlightened mediators and establish free zones, specialised business areas, microenterprises, start-ups, capital risk companies and microfinancing mechanisms.

City–business partnerships are crucial for promoting and implementing innovative projects leading to sustainable development. The Bremen declaration "Business and Municipality: New Partnerships for the Twenty-First Century" adopted in 1997, focused on local, national, regional and international framework conditions for the sustainable development of communities and economies. According to the declaration, businesses ask cities for favourable locations, low taxes and fees, preferential treatment and social recognition as a significant employer. Municipalities seek from businesses employment for citizens, tax revenues and investment in the city. At the crossroads of these requirements, win–win investments and benefits are numerous. The Bremen initiative advocates joint excellence and includes awards to stimulate innovative action (Bremen Municipality 1997).

The 2008 World Economic Forum focused attention on cities in recognition of their vital role in meeting global challenges. The concentration of people and activities and the increasing leadership of city mayors and governors present an opportunity for cities to become a powerful force for global change, particularly in the crucial areas of energy efficiency and climate change.

The SlimCity Initiative, launched in Davos during the 2008 World Economic Forum, is a perfect example of concerted action bringing together mayors, regional governors and the private sector. SlimCity is an exchange between cities and businesses to support action on resource efficiency in policy areas such as energy, water, waste, mobility, planning, health and climate change. SlimCity is also a partnership between the World Economic Forum, the World Bank, the International Energy Agency, ICLEI and the global city network and provides a dynamic environment within which city and business leaders can explore together practical solutions, exchange knowledge and showcase best practices.

Crises can signal the advent of a better world and a green and smart economy can be instrumental in achieving this. Until the day when "green" is not be used to designate ecological processes because everything will be green, green economy initiatives can be very beneficial for counteracting industrial decline and financial crisis. A smart and green economy can both help overcome the crisis and help build a new model of development.

The transition to the low-carbon and postcarbon economy should be a source of competitive advantages. In the UK, the decay of the polluting mining industry was the starting point for a number of green initiatives. A former coalfield in Ollerton, near Nottingham, won the 2005 Enterprising Britain award, which recognises towns and cities that successfully drive forward enterprise. Sherwood Energy Village, supported by the East Midlands Development Agency, is an exemplary model for sustainable growth and development. The former coal mine provided an opportunity for a type of green economy based on renewable energies, while seeking to integrate housing, industrial and commercial development, recreational and leisure facilities and educational resources.

An interesting example of enterprises in Sherwood Energy Village is the company Ground Solutions, which exploits geothermal energy from the earth, as coal mining did in the old days. The transition to higher value has been successful and ground problems met with ground solutions. The village constitutes an environmental economy best practice through the promotion of energy-efficient and sound buildings in pleasant environments.

In times of economic crisis, a dependency on traditional industries is often seen as a major risk, linked to monoeconomy patterns. The magic dust is innovation and smart specialisation. In early 2009, the URBACT network Urban Network for Innovation in Ceramics (UNIC), led by Limoges, explored the possibilities for cities to improve their technological and cultural know-how and branch into long-term, sustainable and unique "paths of innovation". The starting point is that cities with strong cultural and economic roots in the ceramics industry have to find together a way to progress. The responses tested are also relevant for cities with links to other traditional sectors such as textiles and furniture, and their message provides a clear ray of hope in difficult times.

Green tourism is one of the green economy sectors that many cities are interested in to enhance their natural and cultural assets and boost their local economy. A growing green tourism industry in Boston helps to both spark new interest in a heritage-oriented destination and encourage the tourism industry to improve its operations. Boston Green Tourism is helping to attract visitors who seek quality leisure linked to nature and demand environmentally friendly hospitality services. A multifaceted programme was established after consultation with all sectors providing tourism eco-services and promoting Boston as a preeminent destination for eco-tourists. Taking advantage of cities' physical and cultural assets, green tourism activities can be as diverse as the cities promoted.

SMEs are considered to be propellants of innovation in Europe. The rise of a global information economy and the decline of the large factory brought difficulties to regions with traditional industries. At the same time, cities and citizens began to demonstrate increased environmental awareness. International developments favoured transformations in the production system and led to the provision of new environmental goods and services, often offered by smart and flexible new structures and networks.

Green growth creates propitious conditions for the development of SMEs in cities. The essence of urbanisation, the concentration of people and activities, offers a diversity of resources needed by both green entrepreneurs and environment-aware consumers. The features of a city favourable to green SMEs include access to precious human resources, the availability of other high-quality services, complementary producers and a market for their products. Cities can also provide the necessary knowledge basis and advanced services and support for innovative green SMEs. The proximity of human talent and economic actors, enhancing competition and thus stimulating innovation, may also confer benefits, as consumers ask for more sophisticated products and well-being services. Small businesses based on skills, knowledge-driven production and eco-services are particularly well suited to capture urban advantages.

Innovative schemes to support SMEs are often necessary, since they most often lack resources and expertise on environmental performance. BEST, established by the Bradford City Council to assist SMEs, organises the Bradford Business Environment Forum providing local businesses with information on environmental legislation and opportunities, innovative action and best practice. Since 1992, the forum has enabled local networking and has helped businesses work efficiently and make cost savings. The forum has also been responsible for the consultation process for Bradford's Business Local Agenda 21 and has worked with a number of organisations to produce an environmental training scheme for the national network of Business Link organisations. The emphasis is on brokering, helping businesses communicate among themselves and with the regulators.

5.3   Employment, the Cornerstone of Sustainable Cities

The crisis that started in 2008 has developed into an unemployment crisis. As a result, unemployment has already reached a record high of 8.5% in OECD countries, 2 years after a 25-year low of 5.6% in 2007, and is projected to exceed 10% in the EU. In European countries, two thirds of the unemployed population reside in urban areas. Unemployed citizens represent untapped socioeconomic opportunities and drain the potential of cities for sustainable development. Unleashing underused sources of renewal may make more resources available to be invested in sustainable development (OECD 2009a).

Cities are major direct and indirect sources of employment. The advent of the green economy leads to the development of new skills and training requirements to support new labour markets. Cities can create or encourage employment for rehabilitation and renewal works, landscaping, resource, water and waste management, clean transport and fuels, protection of biological diversity and indigenous flora and fauna, heritage preservation and cultural tourism, public safety, building improvements and neighbourhood management. All these activities are important sources of new jobs conducive to sustainable development.

Just 15 years after the collapse of the mining industry in the Sherwood Energy Village, the success is also recognised in terms of generation of green jobs. Until the mid-1990s, 600 people still worked in mining. With the transformation of the area, 1,200 jobs have already been created in healthy environments and 2,000 jobs are to be created soon. It is estimated that, in the UK, the green economy holds the potential for 400,000 jobs.

Sustainability actions have to balance concern for environmentally friendly processes, products and services with care for employment. Well-targeted, integrated environment and employment policy measures should focus not only on quantitative considerations but also on qualitative ones, especially on qualifications, training and employment conditions.

Environmental programmes have created thousands of jobs in the domains of environmental counselling and services, emission prevention, depollution of heavy industries, research and development, the construction industry and resource and waste management. The preparation of Local Agenda 21 plans has been an important source of employment for many governments. The range of activities and the numbers of jobs are expanding. The identification and evaluation of all environmental assets and the eco-audit schemes and advice created multiple needs for training and new curricula.

As direct employers, cities can boost employment creation for sustainable development. The environmental upgrading and management of buildings belonging to or depending on the municipality, such as hospitals and schools, is a mine of new employment in the rehabilitation and sustainable regeneration sector. Furthermore, improvement of social housing can create a pool of jobs. Housing improvements are highly labour intensive. The energy-efficient renewal of social housing in Paris proved to be a great source of quality employment (Delanoë 2008).

The municipality of Stockholm, with 42,500 employees, is the country's largest employer. Furthermore, the city indirectly creates many jobs. The business sector is dynamic and one out of three foreign-owned companies in Sweden is located in the capital city. Education levels are high (51% of Stockholmers have studied at university level compared with 35% nationwide) and average salaries are proportionally higher.

In Malmö, 7,000 jobs have been created in the former shipyard area Kockums, which went through sustainable regeneration and hosts many new activities, notably Malmö University. This figure represents more employment than during the glorious shipyard days. The quality of employment is also significantly higher (Malmö 2008).

Urban renovation offers multiple opportunities for a wide variety of professional and artistic skills. The study of urban regeneration programmes in Portugal highlighted that, for the same budgetary expenditure, twice as many people on average were employed in the rehabilitation sector, far more employment-intensive than new construction. Public works needed for urban renaissance can profit greatly from the social economy, which often provides a popular and informal form of employment, estimated at about 5% of Europe's overall employment. It could also serve green growth and sustainable development.

Green employment contributes to environmental protection and reduction of humanity's carbon and environmental footprint. Green jobs could range from cleaning brownfields to designing cities where citizens can lead fulfilling lives without much impact on the environment. The question if shifting to the green economy results in a net gain or a net loss of jobs has been analysed and the answers vary. It seems that the renewable energy sector is more labour-intensive than the conventional energy sector. The labour market for retrofitting old housing could thrive, whereas research and development promise many job opportunities (UNEP 2008; UNEP/ILO 2008).

In the USA, the 2009 American Recovery and Reinvestment Act suggests that green jobs would help the economy to regain steam, provide workers with decent, well-paid jobs and develop technologies to benefit the environment. The Sustainable South Bronx group is at the forefront of a movement to put low-income and low-skilled workers in "green collar" jobs, involving manual work in environmental sectors. Since 2003, this group has trained 70 former drug addicts, welfare recipients and former convicts for jobs in landscaping, ecological restoration, green roof installation and management of hazardous waste.

A green knowledge-based economy can also create employment for migrants and the disadvantaged. Dublin provides an interesting example of a city with very little influx of migrants and very high unemployment until the early 1990s. The Celtic tiger phenomenon, with an economy growing by more than 6% per year during the period 1995–2005, not only created jobs for the unemployed but attracted many migrant workers to a country much more familiar with emigration. Ten percent of the population of Dublin is composed of migrants, mostly from eastern Europe. And although the economic downturn slowed down new migration, the city can count on a more diversified and better trained labour force for the green economy.

A sustainable community allowance programme inspired by the New Economics Foundation, in the UK, could generate immediate employment at the community level by working with existing organisations to identify neighbourhood needs linked to social issues, local environments, safety, etc. The social return on investment is estimated to be 10 times the amount invested.

Fiscal reform and the introduction of economic instruments may create double dividends, with taxation moving from labour to the environment. Environmental taxes may induce structural changes in industry and favour employment-intensive sectors, especially in labour-rich but resource-poor world regions such as Europe.

5.4 Quality of Life as an Asset for Cities

All cities wish to attract people and capital. The attractiveness of a city can be assessed through the analysis of the waves of people and businesses emigrating from and immigrating to a city. Studies undertaken both at the international and at the European horizon have tried to evaluate the "potential of attraction" of cities for human and financial capital.

The quality of life offered by cities is increasingly regarded as an essential competitiveness factor. The surveys on the quality of life in 270 world cities conducted annually by the international consultancy on human resources Mercer offer interesting insight. Mercer's surveys are based on a range of services which correspond to the typical expenditure of the population. Rankings are based on a score-point index. Cities are ranked against New York, serving as the base city with an index score of 100. The surveys constitute valuable sources of information for governments and multinational companies which have to take into account a series of factors when structuring remuneration packages for their expatriated employees.

The Mercer assessment surveys are based on 39 key indicators addressing political, social, economic and environmental factors. The surveys assess the political and social environment (political stability, safety), economic environment (GDP per capita, inflation, interest rates, banking services), sociocultural environment (censure, limitations on personal freedom), public health and medical departments (health care, water provision and health services, infectious diseases, air pollution), education (level and availability of the schools), public services and transport (electricity, water, public transport, traffic jams), leisure and culture (restaurants, theatres, cinemas, sports), consumer goods (availability and cost of food and consumption articles, and of cars) and housing.

European cities dominate the rankings of locations with the best quality of living, according to the Mercer Quality of Living Survey. In 2008, Zurich was, as in 2007, the highest ranked city, followed jointly by Vienna and Geneva, then Vancouver and Auckland. The highest entry for the USA was Honolulu, occupying the 28th place. The cities with the lowest quality of living ranking were N'Djamena, Khartoum, Brazzaville, and Bangui. Baghdad, ranking 215, retained its position at the bottom of the list (Mercer 2007, 2008).

The survey also identified those cities with the highest personal safety based on internal stability, effectiveness of law enforcement and relationships with other countries. Luxembourg was at the highest position, followed by Bern, Geneva, Helsinki and Zurich, all equally placed at number two. Chicago, Houston and San Francisco were noted among the safest cities in the USA.

The 2009 Mercer Quality of Living Survey revealed that Vienna has overtaken Zurich as the world city with the best quality of life, whereas Geneva retains its third position, and Vancouver and Auckland share fourth place. In general, European cities continue to obtain the highest scores. The top ten places are occupied by German and Swiss cities, including Düsseldorf (6), Munich (7), Frankfurt (8) and Bern (9). In the UK, London ranks at 38, whereas Birmingham and Glasgow are jointly at 56. In the USA, the highest ranking entry is Honolulu at position 29. Singapore (26) is the top-scoring Asian city, followed by Tokyo at 35. Baghdad, ranking 215, remains in the lowest place of the classification (Mercer 2009a).

Many eastern European cities have seen an increase in quality of living. A number of countries which joined the EU in 2004 have experienced consistent improvement with increased stability, rising living standards and greater availability of international consumer goods. Ljubljana, for example, moved up four places to reach 78, whereas Bratislava moved up three places to 88. Zagreb moved up three places to 103.

As a result of the financial crisis that started in 2008, multinationals are looking to review their international assignment policies with a view to cutting costs. Many companies plan to reduce the number of medium to long-term international assignments. The surveys on cost of living are especially useful.

According to Mercer's 2009 Cost of Living Survey, Tokyo (scoring 143.7 points) has knocked Moscow off the top position to become the world's most expensive city. Osaka is in second place and Moscow is in third place. Geneva climbed to fourth position and Hong Kong moved up to fifth. Johannesburg (49.6) has replaced Asunción in Paraguay as the least expensive city in the ranking. The survey covers 143 cities across six continents and measures the comparative cost of over 200 items in each location, including housing, transport, food, clothing, household goods, culture, services and entertainment (Mercer 2009b).

Urban infrastructure has a significant effect on the quality of living experienced by expatriates. Although often taken for granted when functioning to a high standard, a city's infrastructure can generate severe hardship when it is lacking. The 2009 ranking also identifies the cities with the best infrastructure based on electricity supply, water availability, telephone and mail services, public transport provision, traffic congestion and the range of international flight connections. Singapore is at the top of this index, followed by Munich in second place and Copenhagen in third place. The Japanese cities Tsukuba (4) and Yokohama (5) come next, and Düsseldorf and Vancouver share sixth place and Frankfurt occupies eighth place, jointly with London.

Cushman and Wakefield, a global property service provider, has been carrying out surveys into European business cities since 1990. The underlying data are studied independently by the company Taylor Nelson Sofres. Each survey is based on the opinions of the senior management of 500 European companies and according to the criteria considered important at the time of the location decisions. The key factors taken into account by the companies include accessibility to markets, availability of qualified personnel, transport and communications and the cost of living.

The 2009 European Cities Monitor is based on interviews with senior managers and board directors in charge of the location for 500 of Europe's largest companies. The analysis concludes with an overall ranking of European cities considered by the business community to be "best for business" and the "best city in which to locate a business today". In addition to the overall ranking, 34 cities are ranked against a number of criteria, such as quality of life, telecommunications, access to markets, availability and quality of staff, cost of office space and transport links (Cushman and Wakefield 2009).

In its 20th year, the European Cities Monitor (2009) features a special analysis of the performance of cities since 1990. All of the key central and eastern European cities have seen their position improve, especially the ones that benefited from EU accession. Prague and Warsaw are the biggest risers to 21st and 23rd place, respectively, moving up from their positions at the bottom of the ranking in 1990. Warsaw is a financial and political centre which seems to be a good and reliable partner for long-term investments. The rise in Cushman and Wakefield's ranking suggests that the Polish capital is an investor-friendly city and has continued its dynamic growth. Warsaw appears resilient to the global crisis and successfully implements strategic investments, maintaining the investment budget at a record level.

As for recent years and despite the effects of the crisis, London appears to be the leading business and financial centre. It ranks as the number one city in half of the 12 major rankings, including access to markets, availability of qualified staff and international and internal transport links. It scores poorly, however, on the cost of staff, the cost of office space and levels of pollution. Major companies clearly recognise that its critical mass of talent, facilities and infrastructure make the city a compelling location for business. As the host for the 2012 Olympic Games, it will also benefit from a large domestic investment in infrastructure.

Barcelona and Madrid rank fourth and sixth, respectively, against only 11th and 17th in 1990. Both Spanish cities demonstrated their sustainable significant rise over the last 20 years. Both cities are business, culture and leisure centres and cosmopolitan international cities. Barcelona is highly valued among business decision-makers and scores high in almost all the important categories, with exceptional scoring on office availability and value for money. It also consistently ranks first in the quality of life category (Cushman and Wakefield 2009).

Companies were invited to specify the factors likely to influence their business during the next 10 years. The market opportunities offered by the economies in full expansion of China and India constitute the first factor, followed by the performance of the Euroeconomies and, in third place, the USA. One company in six (17%) envisages delocalising or externalising operations during the next 2 years. Eastern Europe seems the favourite destination, whereas China exceeded India as the supreme destination outside Europe. Almost a quarter of the companies interviewed had externalised operations during the previous 3 years.

The state of a city has a strong impact on the state of the citizens' lives and determines the attraction of a city. The Veolia Observatory of the urban ways of life in 2008 aimed at collecting and analysing the citizens' perceptions and aspirations. The survey was based on 8,500 interviews in 14 metropolises on five continents and allowed the collection of comparable data on the ways that citizens feel about and live in the city and their desires for the city. A qualitative study allowed evidence to be collected from groups and live portraits to be obtained which disclose the human face of the cities.

The Veolia Observatory revealed that the city is a source of intense feelings, much more positive (86%) than negative (54%). City residents appreciate the advantages that cities offer them, but also recognise that the price to pay is often rather high. Seventy-five percent of the city residents chose to, rather than are obliged to, live in their city of residence. Among the factors defining quality of life, 36% of the city residents suggested the cost of acceptable living and safety come first, followed by the environment (27%), infrastructures (26%), public transport (24%) and leisure (21%). Concerning their future, many more city residents are optimistic (65%) than anxious (35%). The ideal city should be more open, citizen-friendly, greener and have cleaner air, but should also be more secure, a city with the colours of the blue, of the green and of the white (Veolia Environment 2008).

5.5 The Competitive Edge of Cities

Green parks are the lungs of a city, the places where the natural and human-made environments interact. Maintaining green land often has to resist enormous pressure when the financial development cost of the land is very high. Parks provide space for outdoor recreation, and contribute to the preservation of biodiversity. The amount of green area per inhabitant is a common indicator of living standards in cities, often assessed in conjunction with accessibility patterns. Parks and green spaces are usually treasured places in urban regions and hundreds of civil society organisations are involved in the preservation of green spaces in urban regions.

The lungs of the cities and heavens of biodiversity, green parks are often subjects of heated debates in relation to their preservation. London's biodiversity strategy provides an overview of the city's wildlife and sets out 14 policies to be implemented by the London Biodiversity Partnership, a consortium with the most important organisations working on biodiversity in the UK capital. The city has a wealth of places for wildlife and many organisations run educational and leisure activities for children and the public. The plan includes policies to protect, manage and enhance biodiversity. The Blue Ribbon Network for the River Thames and London's other waterways was set up to establish priorities for the use and management of the water and the shore. The creation of green places is encouraged and the city promotes public access and environmental education.

Some cities are endowed with large forests accessible to all citizens by public transport. At the very heart of the Stockholm metropolitan area, the national city park of Haga-Brunnsviken-Djurgården-Ulriksdal, known as the Stockholm Eco-park, offers an example of an invaluable asset of crucial importance for urban ecology, recreation and biodiversity. The Forest of Soignes, in Brussels, valued as a jewel second only to the sixteenth century Grand Place, accounts for 60% of the region's green space. A protected landscape, the forest is a unique ecosystem. Sixty-five percent of its surface is covered by towering beeches growing on average 5–10 m higher than they do elsewhere in the world. Citizens want the majesty of the forest to be kept intact, each tree standing as a living monument.

The sustainable management of parks is very important for quality of life in cities. The Brussels Institute for Management of the Environment, in control of the 38% of the forest, undertook, when preparing a management plan in 2000, public consultations and canvassed the opinions of 40 ecological associations. A clear majority favoured the preservation of half of the beech forest and this was incorporated into the Management Plan of the Brussels Region. The remaining half would be transformed into a more diversified forest to stimulate biodiversity in accordance with international biodiversity agreements.

Green urban areas can improve the microclimate, favour physical exercise and well-being, promote nature conservation, education and research and provide a higher experience value of the urban environment with the feeling of the changing seasons, colours and odours. Urban forests can help cities meet important environmental and public health challenges, including reducing greenhouse gas emissions, improving air quality and calming crime spots. However, large central parks may not represent the most sustainable solution in cities. Many experts suggest smaller, even "pocket parks" at the level of neighbourhoods and large parks at the peripheries of cities, defining the boundary between the city and the countryside.

Owen (2009) suggests that large parks, including Central Park in New York, may have many of the drawbacks of sprawling suburbs as they function as a barrier to car flows and are not endowed with cultural facilities designed to sustain an unbroken chain of lively human interactions. Much depends on the accessibility conditions, which are impacted by the real and perceived integration of green parks into the urban fabric.

Boston's Urban Tree Canopy Initiative is dedicated to increasing its tree cover by 20% by 2020. By planting approximately 100,000 trees, the Boston Urban Forest Coalition hopes to improve the health and well-being of Boston's residents and visitors and reduce the contribution of the city to climate change. In 2006, the first ever inventory of Boston's urban flora was completed and revealed that Boston has 34,497 street trees, 26,527 of which are in good condition. Overall, Boston features 29% canopy cover, including all trees in parks, private yards, and along streets, but their spatial distribution should be improved.

After the green parks, the lungs of the cities, the grey parks, including science, technology, industry and business parks, technopolises or edge cities, are mushrooming throughout the world and may be seen as the brains in cities. The wave of grey parks started when manufacturing industries, traditional motors of growth, were supplanted by the service sector.

The need for common infrastructures, research and development activities and services is the major reason for the organisation of science parks, some of which evolved into science cities. Teleports, parks providing high-technology infrastructure for businesses that are heavily dependent on telecommunications, have also expanded (Castells and Hall 1994).

The edge cities of the 1990s, especially in USA, developed as new suburban cities located closer to major highways and hosted glistening office towers and huge retail complexes (Garreau 1991). A new generation of science, technology and business parks exhibits sustainability principles by turning brownfields into healthy areas and by connecting them to major public transport exchanges.

From Tokyo to Barcelona, and from Malaga to Athens, former industrial estates have been reshaped and enriched with infrastructure enabling them to attract grey investments. The Technopolis of Bari comprises a research centre, a training centre and international schools and services. Profits derive from the provision of services to businesses and the public administration. The parks are designed to respond to demands for innovation and also to create a new demand for innovation in a high-quality smart and healthy environment.

The transformation of brownfields into greenfields has been a decisive step in converting urban liabilities into sustainable assets. The conversion of old industrial sites into clean factories holds much potential. Abandoned and often contaminated industrial sites, landfills and mines, usually connected to infrastructures and economies, are being redeveloped to host clean manufacturing. The US Environmental Protection Agency and the National Renewable Energy Laboratory have mapped almost 4,100 contaminated sites which could be utilised for wind, solar or biomass development. Fully developed, they could produce more than the country's total electricity consumption.

Sustainable grey parks often provide interesting cases of public–private partnerships for turning deprived areas into spaces bringing environmental and economic benefits. Stockley Park, a former derelict rubbish tip, within the green belt to the west of London, is an inspiring example. A partnership between the developer, the local authority and the university created an international business park and public parkland including recreational facilities. In exchange for the right to construct the business park over 36 ha, the developer guaranteed the development of the whole site of 140 ha, removal of groundwater pollution, environmental enhancement and landscaping. At all stages of the creation of Stockley Park, local residents were involved in the process through extensive community consultation.

In Germany, the IBA Emscher Park has been an extraordinary and monumental project of urban and regional development and ecological restructuring within the northern Ruhr district. During German reunification, European experts worked together with the cities and industries of the Emscher region to transform coal mining settlements and the industrial cathedrals of the past into sound modern communities with healthy housing, attractive spaces and high-quality locations for industry and services.

The landscaping of the Emscher area, the ecological restructuring and the protection of the green environment and the water environment were key elements. Art has also played an important role and artistic interventions among the Ruhr and Ems rivers had a decisive contribution to the new identity of the region. On the traces of Zollverein, the largest mine, famous architects proposed a cultural sense for the new life on the site. The intercommunality model developed after the enhancement of the quality of the area has been a lesson in urban management (IBA 1999).

Watercolour 6
Helsinki: A Competitive and Attractive City?

Helsinki: A competitive and attractive city?

Chapter 6
Social Justice and Solidarity in Cities

This chapter sheds light on the social capital of cities, as concerts of traditions, hopes and wills. Cities are schools for respecting differences and learning life in society. Social justice is a prime criterion for judging urban sustainability. Unequal distribution of wealth has draining effects on the vitality of cities and it is a source of both unsustainable lifestyles and obstacles to cultural change. Equity and social justice, public health and well-being, eco-responsible housing and lifestyles, safety and solidarity touch the very heart of cities and impact their potential for sustainability.

Distressed areas, the back scene of most urban theatres, the places where most disadvantaged and excluded citizens congregate, have much potential for sustainable development. Citizens can play a major role in creating vital urban spaces and fulfilling neighbourhoods out of degraded spaces. Participation of youths and women in projects can especially extend the limits of the possible. Education remains a major imperative.

6.1 From Distressed to Fulfilling Cities

A strong human and social capital is a characteristic of the sustainable city with a human face and a cohesive society, which has the capacity to withstand threats. Social justice is a precondition for the creation of sustainable wealth. Fair distribution of resources constitutes a fundamental ethos of the social architecture of cities, which does not accept being polysegmented or a city of "forced solidarity". Unequal distribution of wealth has draining effects on the vitality of cities and it is a source of both unsustainable lifestyles and obstacles to cultural change.

The main social challenges for European cities include social justice and solidarity, high-quality employment, the ageings of the population, the recognition of differences and the enhancement of opportunities linked to the urban diversity. Longer healthy lives are a great chance for citizens and cities wishing to address the expectations of an older population. The role of equity and social justice in shaping sustainability processes is unparalleled. David Harvey reminds us that "there is nothing more unequal than the equal treatment of unequals."

V.P. Mega, *Sustainable Cities for the Third Millennium: The Odyssey of Urban Excellence*, DOI 10.1007/978-1-4419-6037-5_6,
© Springer Science+Business Media, LLC 2010

Welcoming migrants is a challenge for many cities of the more developed countries. Net migration to the EU tripled between 1994 and 2003 and reached two million. Cities often experience tensions between spatial proximity and cultural distance, linked to the coexistence of migrants with other social groups. More selective and better integrated immigration would be more beneficial. Education, housing, health, sports and politics are the sectors with the highest potential for the integration of migrants in the city (EC 2008a).

Preventing and fighting social exclusion is a major issue for cities suffering from various forms of schizophrenia with multiple phenomena of urban distress. Even in the most prosperous European cities, there are urban islands where environmental degradation and social exclusion reinforce each other. More or less extended zones in run-down city centres or chaotic peripheral zones invite the tearing down of walls and the creation of true urban spaces and life.

Urban decay indicates that the capacity of urban systems to innovate and drive change is overstretched and the urban cell renewal slows down. Symptoms include invisible features and visible signs. Distressed neighbourhoods suffer functional impoverishment, with destitute housing and insufficient equipment and facilities, delinquency and crime, high unemployment, low mobility and little access to information, education and training. Very often, transport infrastructures, whether operating or disused, break continuity and further isolate distressed areas from vibrant urban centres.

Urban distress is an obstacle to sustainable prosperity. A quartered city mirrors multiple divisions. Distress often results in a dichotomy between spaces and functions. The OECD research programme on distressed urban areas shed light on both quantitative and qualitative characteristics of such areas and proposed actions for prevention and regeneration.

Social cohesion is linked to the capacity of a society to endure threats. It is a dynamic and multidimensional process, interconnected to labour markets, housing, health, education, welfare systems and citizenship. The main factors at the origin of the processes of social exclusion include globalisation, economic restructuring, competition between companies, cities, regions and nations, and the restructuring of welfare states.

Cities are the homes of the poor and the homeless, who make up 16% of the European population (EC 2008a). It is estimated that 7% of Europeans are persistently poor. Poverty-trap conditions of low income and welfare-dependent economic structures can exacerbate exclusion. Increased financial pressures, in a complex and fragmented institutional environment, have to be addressed through the horizontal and vertical integration of decision-making systems and the optimisation of the contribution and commitment of the public, private and social economy sectors (Parkinson 1998).

To preserve their social capital, cities must regenerate their spatial and social fabric (Galbraith 1996). They must offer all citizens access to information, education and training, adequate housing and noble public spaces, social services and the possibility to participate in codesigning the future of the city.

Cultural diversity is a resource for the dynamism and growth of a city. Differences and conflicts involving migrants or minority groups can, if left unmanaged, undermine the city's sense of community and identity and weaken its ability to adapt to change and attract investment. If cities manage diversity properly, they can benefit hugely from the potential of migrants and minorities for entrepreneurship and innovation. For this, they should review the range of institutions, services and policies to create the appropriate conditions and governance structures.

Multicultural cities hosting diverse cultures have to create the conditions for people not simply to coexist peacefully but to engage in a fruitful dialogue. The Intercultural City programme, jointly promoted by the Council of Europe and the European Commission, tries to enable cities to address pertinently the challenge of cultural diversity. Through the intercultural city policy innovation network, a learning community of cities, politicians, practitioners, academics and members of society, the project facilitates mutual mentoring and exchange between cities. The activities of the network are designed in a way which allows a broad range of actors, city officials, service providers, professionals and civil society organisations to be involved in the process of building an intercultural vision and strategy for the cities, reviewing polices through the "intercultural lens" and developing skills and assets.

Lyon "intra muros," promoted as a city of confluences, is a key actor in this programme. Since ancient times, it has been a welcome land for migrants from the Mediterranean, Asian, European and African countries. The "small Lyon" has its "migrant neighbourhoods," nine official districts, comprising vast housing areas in the western outskirts of the city and a fascinating inner quarter, the "Guillotière," a lively, colourful, cosmopolitan neighbourhood known as the gate of the city where newcomers traditionally arrive. Lyon was also the first city to sign the "Diversity Charter" for non-discrimination in employment.

Interfaith cultural centres and events had much impact in many disadvantaged suburbs where minorities coexist. Interreligious open days can help the dialogue among populations that do not usually communicate among themselves. Curitiba, the ecological capital of Brazil, initiated a project called the "Lighthouses of Knowledge." These Lighthouses include free educational centres, libraries, Internet access and other cultural resources. Job training, social welfare and training programmes are coordinated and often supply labour to improve public services.

6.2   Youths and Women: The Actors of Urban Solidarity and the Stake of Education

All citizens have equal duties and rights to the city, but the symbolic importance of some social groups is incomparable. Youths, the future of society, often constitute the most vulnerable part of society and the most acutely affected by economic crises and unemployment. Underprivileged youths often growing up in difficult neighbourhoods or, even worse, living on the streets without prospects for the future, may act violently against the society that has neglected them. The integration process encompasses simultaneous interventions in education, employment, housing, family structures and services. Dropping out of school is considered as a first step towards exclusion. The educational system has to impart respect for each other and for the community, in addition to knowledge.

Children, often "invisible in statistics", are the youths and citizens of tomorrow. The African proverb "It takes a village to raise a child" is also very valuable for sustainability. Hundreds of municipalities have established municipal councils for children and involve young citizens in discussions about the possible urban futures. In Finland, the "Children as Urban Planners" project in Kitee aimed at educating active citizens in environmental awareness and responsibility for their built and natural environment. In Helsinki, hundreds of schoolchildren studied the urban history of their capital city and then redesigned the historic centre of Helsinki (EFLWC 1996b; OECD 2009b).

Children are at the very heart of sustainable development. The well-being of children is a litmus test for the present and future of society as a whole. For children to do well, their families have to do well. And for families to do well, their community must do well and find its own way to sustainable development.

In the Philippines, the Council for the Welfare of Children initiated the Presidential Award for child-friendly municipalities and cities. The award is considered as a self-assessment mechanism of the level of child-friendliness of a particular city and municipality and constitutes recognition of cities which create child-friendly environments and uphold the rights of children regarding protection, development and participation. The Council for the Welfare of Children oversees the process of screening, evaluation, validation and selection of winners of the programme and project accomplishments benefiting children and their families.

In New York, the Harlem Children's Zone Project offers an example of a holistic approach to rebuilding a community so that its children can develop fulfilling lives through education and access to the labour market. The goal is to create in the neighbourhood an enriching human and social environment which can act as a counterweight to "the street" and help children as early in their lives as possible. In January 2007, the Children's Zone Project expanded its comprehensive system of programmes to nearly 100 blocks of central Harlem. The project includes in-school, after-school, social services, health education and community-building programmes for children of every age. The project also works to reweave the social fabric of Harlem, which has been torn apart by poverty, crime and drugs. The Harlem Children's Zone Project served as a model for the Promise Neighbourhoods announced in 2009 by the American president (Obama 2009).

Human capital is considered to be the most important in cities, and universities are often regarded as the single most important asset. Education and research can instil a culture of continuous improvement. Athletic and sporting activities favour "healthy minds in healthy bodies" and social integration and are important both for the human and the social capital of a city. Empowerment has become ethically correct and practically advantageous. It helps develop leadership at every level.

Science weeks include structured pedagogical actions for youths to initiate smarter consumption lifestyles. Eco-schools, museums and eco-stations organise special events to raise awareness among young people and engage them in creative action towards sustainability. In Hamburg, the fifty–fifty project involves all schools in the city, committed in radically reducing energy and water consumption and successfully implementing climate protection plans. The Hamburg environment centre organises children's research workshops with learning stations for solar energy, fuel cells, heating systems, etc. During the science fair, all schools participate in Climate Action Day.

In Amsterdam, the Energy Survival programme offers children the opportunity to learn the importance of sustainable energy through an education discovery expedition. Young people emphatically take command of actions for better energy use. In 2007, seven primary schools took part in the programme, also comprising an energy survival day (Amsterdam Climate Office 2008).

Many innovative urban programmes are centred on the action of the parents and in particular of women, bearers of a special sensitiveness. One out of every two "homo urbanus" is a woman, the "other half of the sky" (Odysseus Elytis). Women are overrepresented among the poor and are usually closer to everyday life and therefore able to bring a reality check in validating policies. Beyond the divides of cultural incomprehension, the conscience of being part both of a community and of the world is most often cultivated by women. Wangari Muta Maathai, the founder of Africa's Green Belt Movement is a symbol of persistent struggle for environmental improvement, democracy and human and equal rights. It has been proposed that the single most productive investment for sustainable development on the planet is considered to be the education of women in Africa (Harvard University 1994).

Gender mainstreaming is considered to be an important step forward, compared with traditional equal opportunities policies, often perceived as a social policy field for preventing and fighting discrimination towards women. Gender mainstreaming has to be integrated in all areas of public and private decision-making, to achieve the goal of equality. According to the Council of Europe, gender mainstreaming entails the reorganisation, development and evaluation of policy processes, so that a gender equality perspective is incorporated in all policies at all levels and at all stages.

The Gender Gap Index introduced by the World Economic Forum is an interesting measure of the way that countries and cities divide their resources and opportunities among their male and female populations. Europe has eight out of the ten best performers, with Sweden, Norway, Finland and Iceland as the top rankers. All countries in the top 20 made progress relative to their previous-year scores. Latvia and Lithuania made the biggest advances among the top 20, gaining six and seven places, respectively, driven by smaller gender gaps in labour force participation and wages (WEF 2007).

Ensuring that a gender equality perspective is incorporated in all urban policies means that all proposals and decisions related to all urban activities and policies must be considered from the angles of both genders. A detailed analysis of the gender implications sometimes reveals that different measures are required for women and men and at different levels. Time management and urban security have been highlighted as domains where measures may differ considerably. "The Cities' Manifesto for Safety and Democracy" shed light on violence perpetrated against women and suggested that urban safety measures integrate gender-specific approaches (EFUS 2006).

Moving towards gender-conscious cities engages the active participation of women as a condition enabling progress towards sustainable development. Women, notoriously rooted in concrete realities, could bring an element of balance in urban policies. Women politicians are often more sensitive to schemes reconciling work and family life. Many innovative urban projects are due to the enthusiasm and effort of women architects, scientists and politicians. The Athena project in Sweden and the Frauen project in Vienna-Donnaufeld are exemplary housing programmes, which aim at creating housing units conceived by female architects and responding to women's needs.

6.3 Sustainable Housing: The Living Cells of the Cities

Sustainable cities need sound living cells under perpetual renewal. The deterioration of housing environments may have dramatic consequences for the well-being of cities and citizens. Social public housing has often scarred the urban fringe with dysfunctional suburban areas that drain citizens of respect and responsibility.

In many urban peripheries, a new human face has to replace the alienating towers set in landscapes of dreary desolation. Housing estates built quickly and cheaply after the Second World War, as if they were to house interchangeable people, have no meaning in the search for sustainable development. Concerning housing models for a diverse sustainable society, one type fits none.

The concept of Choice Neighbourhoods announced by the American president recognises that different communities have different housing needs. Homogeneous and monolithic social housing projects that too often trap residents in a cycle of poverty and marginalisation, cannot support vibrant communities and enhance opportunity for residents and businesses (Obama 2009).

In many European cities, housing has begun to respond to the evolving requirements for sustainable development (eco-housing, eco-management of housing units) and the social evolutions (ageing, smaller households, etc.) Vibrant local communities, with corporate neighbourhood space and responsive management, are replacing neglected or deprived neighbourhoods.

Renewal schemes for residential areas provide many interesting lessons. In city centres, soft renewal allows inhabitants to remain in place and avail themselves of a range of resident-friendly measures. The soft urban renewal in Vienna included block improvement schemes, enhancement of public spaces and ecological measures. The social criteria insisted on avoiding segregation and on involving ageing inhabitants in smart renewal management.

Just 20 years after their construction, housing blocks have been demolished and reconstructed with new aims and principles. The Mascagni development, in Reggio Emilia, a multifunctional urban space created from a rigid series of anonymous buildings, bears witness to a purposeful interconnection between old and new housing units, with integrated public services and local businesses. In Hällefors, in Sweden, the transformation of the housing area Klockarskogen, through selective demolition, rehabilitation and reconstruction, provided multiple opportunities. The local economy has been reinvigorated, skills were generated and a sculpture park has been created (EFILWC 1996b).

Government initiatives, such as the City Challenge in the UK, played a decisive role in kick-starting innovations in housing regeneration during the waning years of the twentieth century. An exemplary renewal project is that of Holly Street Estate, constructed during the 1960s and 1970s as a series of blocks, replacing the traditional two-storey east London houses, and notorious two decades later for its state of deprivation, crime and delinquency. The local authority recognised that the only way to deal with the situation was the demolition and reconstruction of the housing estate. The renewal project was initiated in response to the UK Government's Comprehensive Estates Initiative and tried to break the cycle of welfare dependency and poverty. Pleasant Victorian-style houses have replaced the tower blocks, generating greater home identification (Mega 2005).

Sustainable housing expands through intelligent resource-conscious and eco-conscious neighbourhoods and sensible land use. Some inspired experiences highlight that social housing should not be synonymous with poor architecture and neglected environments. The Solar Village, created by the Social Housing Association in Greece, as a residential village for low-income households, is a good example of design for healthy environments and personal identification. The architectural care and the social features merit special attention. The ecological design and planning of the area tried to make the best out of the physical parameters (Tombazis 2007).

The passionate involvement of citizens can often compensate for scarce resources. In Finland, the strength of the Top Toijla project is due to the commitment of the mobilised inhabitants for the improvement of the Rautala housing area. Ambitious renewal has been achieved with a modest budget. A community theatre has been created to identify and solve problems and nourish visions and actions.

Like many cities in the UK, Glasgow witnessed the construction of high-rise housing blocks in the 1960s. Towers were built to replace the decaying buildings, originally hosting immigrant workers. The massive demand soon outstripped new supply and many tenements became overcrowded and unsanitary. Many degenerated into the infamous Glasgow slums, such as the Gorbals. Efforts to improve the housing situation, most successfully with the urban renewal initiatives entailed the comprehensive demolition of slum tenement areas, the development of new towns on the periphery of the city and the construction of tower blocks. The Glasgow Housing Association, a private not-for-profit company created by the Scottish Executive for the purpose of owning and managing Glasgow's social housing stock, took ownership of the stock in 2003 and began a clearance and demolition programme to create a new face and fabric.

In many cities, the dismantling of large concrete blocks of housing estates has to serve the reconstruction of the neighbourhoods as social entities, not just as building blocks, through the provision of cultural and educational activities. In the Duchere neighbourhood of Lyon, the partial deconstruction of giant housing blocks and the construction of smaller housing units and public spaces tried to attract middle-class families and restore the social fabric.

Residential eco-villages, including clustered housing, a number of ecological features and public common spaces and services are gaining ground. The Danish cohousing concept offers an innovative approach, reconciling the need for new housing with the demand for sustainable development. There are many decades of cohousing communities in Denmark, each comprising 20–50 households. They form self-sustained communities, consisting of detached and owner-occupied houses, each of them designed by the owner. Public services typically include a communal house, a playground, an organic garden for biocultivated local products and wind turbines for electricity generation.

Improving the living conditions and health of people at risk of homelessness is a noble aim for most cities. Homelessness is a complex phenomenon and is generally more than the lack of accommodation. Many different needs have to be considered and support has to be targeted and tailored to the individual circumstances. As these needs vary much within and between cities, local approaches differ, but the common denominator lies with the multidimensionality of responses.

In the Netherlands, the National Strategic Plan for Social Relief adopted in 2006 was a breakthrough in integral care for the four largest cities. In Utrecht, with 1,250 homeless people and 3,200 people at risk of homelessness, the plan was implemented by a partnership, expanding the coalition of the city with Agis, the regional care agency, the care and shelter services, the police, the criminal justice system, social housing companies and public and NGO services. The shared aim of this partnership is to provide all homeless people with a personal pathway to independence. An improved gateway to integrated services including care, housing, income and day activities has been launched and there are three subprogrammes focusing on the themes of access, prevention and rehabilitation. In the past, Utrecht had invested heavily in supporting 600 drug addicts to stop sleeping rough. The programme, initiated in 2000, dramatically reduced the targeted population and services have been reduced accordingly (Eurocities 2009).

6.4 Public Health, Well-Being and Safety

Public health is not merely the absence of epidemics or disease, but it is a state of complete public physical, mental and social well-being. Public health and urban wellness are highly interconnected and there is no single fact or policy concerning the urban environment that does not have a direct or indirect impact on health. A healthy city is a city committed to having public health high on its political agenda.

Investing in public health reinforces the immune system of cities and citizens. The WHO project on "healthy cities," launched in 1986, has stimulated important actions as regards public health and the introduction of rapid alert systems. The WHO Healthy Cities programme engages local governments in health development through a process of political commitment, institutional change, capacity building, partnership-based planning and innovative projects. It promotes comprehensive and systematic policy and planning with special emphasis on health inequalities and urban poverty, the needs of vulnerable groups, participatory governance and the social, economic and environmental determinants of health. It also strives to include health considerations in sustainable regeneration and urban development (WHO 2003, 2008).

The WHO European Healthy Cities Network consists of a network of cities around Europe that are committed to health and sustainable development. Over 1,200 cities and towns from over 30 countries in the WHO European Region are healthy cities. Each 5-year phase focuses on a number of core priority themes and is launched with a political declaration and a set of strategic goals. Phase V (2009–2013) has the overarching goal of health and health equity in all local policies. The three pivotal themes include caring and supportive environments, healthy living and healthy urban design.

Public safety is another major component of wellness in cities and a challenge for governments at all levels. Traffic accidents and delinquency seriously affect the quality of life. One hundred and twenty citizens die every day on the roads of the EU. Road safety can considerably be increased through improved vehicle design, road planning and monitoring, together with regulations on drinking and driving and the wearing of seat belts. The EU is striving to cut at least by half the heavy human road toll.

Crime, one of the factors that most affects urban liveability, is in some cases in a linear relationship with unemployment. It also seems linked to the size of human settlements. Many cities have declared zero tolerance to delinquency and national plans assist the setting up of innovative direct or indirect crime prevention plans. Tele- and videosurveillance systems can help reduce crime but they are not always accepted by the population. In Berlin, such systems were accused of reminding the population of the methods used by the Stasi during the Cold War. In London, even though such systems ensure good coverage, crime was only reduced by 3%.

Danish cities prepared action plans targeting the involvement of residents and the strengthening of area consciousness for the creation of safer environments. Guardian angels and street negotiators in Belfast make the city more human. Citizens engaged in improving the everyday quality of life constitute voluntary safety chains in the neighbourhoods of Barcelona.

The European Crime Prevention Award aims to recompense the best crime prevention project. The award, launched in 1997, was inspired by the Roethof Award, a Dutch crime prevention award instituted in 1987. The 2008 award was given to a UK project focusing on a socially deprived neighbourhood of Preston, a city in the northwest of England, which has long suffered from high crime. The estate was in an extreme state of deterioration and in 2003 was placed in the top 5% of deprived communities in the UK. The community spirit was very poor, with little citizen involvement and the environment had supported a gang culture on the streets, drug dealing and a range of other criminality and antisocial behaviour.

In 2005 a new neighbourhood policing team was established and, using widely recognised approaches to partnership and an accepted problem-solving framework, it conducted high-level scanning of the area, followed by a detailed analysis of the problems. One of the agreed objectives was to reduce crime by 15%. The response phase ran through 2006 using a combination of enforcement activity and situational and social crime prevention approaches. This proved to be successful and all the objectives were met or surpassed. The evaluation process included comparison with a control area and measurement of possible displacement to neighbouring estates.

The European Forum for Urban Safety, representing 300 local authorities committed to active policies to fight discrimination and the fear of crime, has promoted the Saragossa Manifesto on "Security, Democracy and Cities," adopted by representatives of several hundred European cities, who met in Saragossa in November 2006, together with representatives of African, North American and Latin American cities.

In 2000, the European Forum for Urban Safety had already promoted the "Cities' Manifesto for Safety and Democracy" expressing the firm will for safe cities, defined as vital places of harmonious development, immune to insecurity and fearfulness, violence, fanaticism and terrorism. Cities preventing crime and terror are contributing to making the EU a space of freedom, safety and justice. The exchange of experiences and cooperation are judged essential for guaranteeing the legitimate right to safety (EFUS 2006, 2007).

The manifesto underlines that safety is an essential public good and interest. Effective integrated policies should be conceived and implemented and fight both the effects and the causes of crime, such as social exclusion, discrimination and socioeconomic inequalities. Cities and local governments should prepare local safety plans integrating prevention of and fighting against organised crime and human trafficking.

Art for all on the street has often been used as a tool for expressing societal discontent and rejection. Graffiti attacks, beyond any form of artistic expression, have been the postmodern way of attacking public and private property. The most frequent targets of aggression include public transport sites and equipment. The repair cost, including the cost of removing the graffiti, is very high for local authorities and property owners. The enterprise for the coordination of public transport in Paris set up a specific service for the prevention of graffiti attacks through research on the attackers and more efficient ways of damage prevention.

An innovative approach to preventing and addressing graffiti in public spaces was developed in Maastricht in the early 1990s when the city was one of the theatres of European integration. The project included enhanced means to identify the offenders, education programmes to improve the skills of the graffiti artists and an antigraffiti bus with previously unemployed young people, including graffiti attackers, specialised in cleaning the public spaces. The city made a wall available to all citizens wishing to express themselves through the art of graffiti. The project proved very successful and led to a dramatic decrease of graffiti on the walls of the city. The identification of wrongdoers and the conditional or alternative punishment had a noticeable effect on preventing recidivism. Moreover, some former offenders have been trained in removing graffiti from public spaces, whereas others became creative graffiti artists.

6.5   Urban Peripheries: From Tensions to Hope

At the back scene of most urban theatres, there are sprawling, formless and disconnected peripheries, which challenge the very notion of the city. They have been developed like "drops of oil" or "blots of ink", without proper extension plans endowing the areas with adequate technical, social and civic infrastructure. This model of growth is not unrelated to the conditions in which industrial production developed during the last century. According to Jean-Paul Sartre, "the developing world starts at the European urban peripheries." The periphery is a zone of great uncertainty and tensions, where people "do not know if they are in or out" (Touraine 1997).

Integrating marginalised peripheries in the urban fabric is a noble but hard task and aim. Many European and national programmes try to create multicultural and civilised places deserving pride and attachment out of anonymous and inhuman suburban islands. Living in the periphery should be seen as a chance and not as a curse (DIV 2004). In Vienna, the Urban Gürtel Plus project aimed at improving living and income conditions in the western Gürtel area, where more than a third of the population are immigrants. The revitalisation of the local economic structures, the introduction of one-stop services and the creation of new jobs have been considered as essential ingredients.

The concept of urban villages, reconciling the intensity of a city with the intimacy of a village, has provoked many debates. Integrating and balancing public interaction and private environments can be very stimulating for politicians, designers and citizens. A true city should grow by meaningful multiplication of its vital cells, its neighbourhoods, and not by overexpansion of its central core. Each neighbourhood should develop its specific unique identity, the result of osmosis between the people and the places. Polycentric cities reflect better the essence of cities as a mosaic of unique places and events which create a potentially recognisable identity. Urban functions and services necessary for ensuring prosperity and quality of life should be found within every urban quarter (Neal 2003).

Peripheral districts often seen as problematic are mines of new opportunities. The Helsinki-Vantaa project, funded under the European Commission Urban Initiative, addressed the problems of the eastern periphery of the Helsinki metropolis, which hosts the highest number of ethnic minorities in Greater Helsinki and is threatened by long-term unemployment, reliance on welfare benefits, illness and crime. It is however, a periphery with an exceptional recreational potential owing to its abundant green spaces, well-organised public transport system, strong sense of community and good prospects for affordable housing. The Urban project tackled the situation on two major fronts. On the business and employment creation front, it was crucial to improve the level of services, safeguard existing jobs and create meaningful new ones. On the social integration and citizen participation front, it was deemed vital to promote capacity building and community development and to support families with problems, immigrants, long-term unemployed and drug addicts.

The transformation of "sensitive" peripheries in high-quality sociocultural and environmental spaces has undeniable effects on growth and employment. Turin is an interesting example. The city plan, adopted in 1993 after 6 years of preparation, introduced a transparent partnership with private investors. Transparency has been a key element, a very important element after the effects of "Tangentopoli." The Lingotto peripheral district became the vanguard of a massive transformation. The emblem of Fiat, the area saw the largest modern factory in Europe, inaugurated in the 1920s. The 500-m-long, five-floor-high building was the first example of modular design in reinforced concrete, a temple of mass production.

The end of production, in 1983, was the starting point for a new era. The company wished to transform the area into a multipurpose space and exhibition centre open to the city and the world. The factory had such symbolic importance that it was essential to grant it a new breath of life. An international architecture competition took place and all proposals were exhibited on the site. A mixed company with the participation of the city of Turin had the task of creating and developing the multipurpose space, including a conference centre, an exhibition hall, an auditorium and a commercial gallery. Renzo Piano's project succeeded in radically transforming Lingotto without betraying the spirit of the place and by making the architectural monumentality more luminous and accessible.

In France, the city policy demonstrated worthy efforts for the revitalisation of deprived suburban areas (DIV 1995, 2009). Unemployment in these areas is more than 40%. In 2003, the three major objectives included urban renewal, support to local actors and solidarity actions. Typical measures included a second chance offered to indebted households and the creation of tax-free zones to attract enterprises. Special contests entitled "Talents de Cités" reward successful initiatives by young entrepreneurs (DIV 2003).

In 2009, the plan "Hope for Suburbs" focused on more than 200 sensitive suburbs, selected as priority action places and the subjects of measures to radically improve housing, environments and employment conditions. Outstanding measures included special care for the youth, coaching for young people looking for a job, a range of traineeships and public–private partnerships to involve the business world and facilitate the transition of young people from education to working life.

At the top parade of difficult peripheries, Shankill Road, in the then divided city of Belfast, in Northern Ireland, is a reference for the renewal of most problematic districts often located in cities in industrial decline. A prosperous industrial centre at the beginning of the twentieth century thanks to the linen, tobacco, rope-making and shipbuilding industries, Belfast had a whole mosaic of working districts, distinguishable between Catholic and Protestant neighbourhoods. The industrial decline and the disruption, conflict, and destruction, called "the Troubles", from 1968 were decisive in worsening the already miserable living conditions in these areas. The suburbs in the north and the west of the city, physically separated by demarcation walls, became the theatre of discord between Catholic and Protestant citizens.

The regeneration of Shankill Road, a Protestant district located between the west and the north of Belfast, offers unparalleled lessons. With the aid of the initiative "Making Belfast Work", a local development agency was created and proposed a comprehensive plan. A cooperation process was set up and organised, in 1995, a community planning weekend entitled "Planning for the Real." The event mobilised 600 people, in particular young people from the district, around priority action workshops and led to the Hankel strategy. The strategy was implemented through a first Urban programme of 14 million ecus, shared with a neighbouring Catholic district. The programme is centred on the action of the parents and in particular of women, considered as carriers of "a new generation of the possible".

Watercolour 7
Prague: A City to Exercise Social Solidarity

Prague, a city to, exercise urban solidarity

Chapter 7
The Cities of Sciences, Culture and the Arts

"Everyone has the right freely to participate in the cultural life of the community, to enjoy the arts and to share in scientific advancement and its benefits," Universal Declaration of Human Rights, Article 27.

"The arts and scientific research shall be free of constraint. Academic freedom shall be respected," The Charter of Fundamental Rights of the European Union, Article 11.

This chapter examines the role of science, culture and the arts in urban sustainability. Knowledge cities and local helices invest in innovation and provide valuable models for targeting investment in science and technology in order for them to become word centres of excellence. Heritage and culture define urban identity and enrich urban capital. Citizens project their hopes and desires into the urban reality and legend, whereas arts in the city represent the ultimate expression of collective intelligence. Cities are beehives of cross-fertilised creativity, the only places where people and resources congregate at a point beyond which synergetic effects become more important than simply additive ones.

7.1 Cities as Beehives of Creativity, the Spark of Excellence and Change

Cities have often flown with the winds of change and shaken the established order. They have brought to the world new concepts by mobilising their unique resources of ideas, enthusiasm, commitment and labour. They are places pulsing with energy and vibrancy, where creativity concentrates and cross-fertilises and human inventiveness gets an extraordinary poignancy. Peter Hall suggests that "innovative cities at their zenith, Athens, Florence, London, Weimar, Berlin, were cities in transition, out of the known, into new and still unknown modes of organisation" (ACDHRD 1995), cities tied to the future and not to the past, despite their often millenary civilisations. Sustainable cities are in a constant state of renewal, able to compose a better urban future through the creativity of their entire population.

Traditional and well-established cities often resist innovation. Discipline, hierarchy and conformity are the enemies of change and resistance increases when

V.P. Mega, *Sustainable Cities for the Third Millennium: The Odyssey of Urban Excellence*, DOI 10.1007/978-1-4419-6037-5_7,
© Springer Science+Business Media, LLC 2010

innovations touch the core interests or boundaries of institutions. Vested interests are the main obstacles to innovation. Local authorities have to challenge hierarchical structures, collaborations, intellectual property management and talent enablement. The most difficult, but potentially most effective, innovation is often to halt established unsustainable practices.

Coalitions that place people at the centre of a genuine forward-looking sustainable development strategy seem to have an unparalleled potential for a powerful transition. Charismatic leaders, experts, citizens and workers are all potential bearers, initiators, generators or supporters of innovations. A common problem or a shared concern often ferments the ground for the coalition. Anticipation of trouble might be decisive. Alliances based on agreement, mediation, political manoeuvring and negotiation can best direct the wave of the future. Mediation can be critical in certain cultural settings as a face-saving measure. Making compromises and reaching an agreement at an early stage is much preferred to tiresome, lengthy, costly and hostility-engendering processes of conflict and arbitration (MIT 1997).

From a new idea to its grafting into a mainstream policy, the birth, growth and death of an innovation depend on a city's creative assets and their mobilisation towards sustainable development. Innovation requires commitment and enthusiasm from the actor(s) that conceive(s) it, confidence that the innovation represents a plausible option and willingness to accept responsibility. It also needs intelligent, efficient and effective coalitions. Responsibility has to be shared among all actors. Nurturing creativity can be contagious and create a climate for continuously mobilising creative potential and stimulating innovation.

Creating new value through innovation has always been an imperative for business. Experts suggest that the future of innovation depends on innovation talent, collaborative innovation and innovation clusters. Cities can influence all of them and provide incentives for innovation to flourish. The development paths of innovation clusters highlight that Tokyo, Ottawa, Sydney and Bristol are the origin of an important number of patents granted by the US Patent and Trademark Office (World Economic Forum 2008).

The Silicon Valleys of the world and European high-technology hubs such as Medicon Valley in Sweden, Cambridge in the UK, Baden-Württemberg in Germany and Catalonia in Spain demonstrate that places are not interchangeable. Some regions have been successful in promoting the geographical concentration of research and innovation actors interacting in clusters. Clustering is local networking, with constituent parts developing strong synergetic links. Clusters bring together academic and research institutions, innovative enterprises and business incubators, local governments and support agencies which mutually reinforce regional receptiveness for the knowledge economy. Their structures embody a knowledge base and an enabling infrastructure, as well as an auspicious socioeconomic and cultural environment. Links may involve a broad range of interactions and exchanges including knowledge transfer, financial transactions or, simply, increased personal contacts. Knowledge spillovers are among the most important by-products.

Smart clustering involves multidisciplinary, plurisectoral linkages and organisations in search of excellence in domains reflecting regional diversity and identity.

Knowledge polygons involve actors who produce, diffuse and enhance knowledge, such as universities, research centres, science and technology parks and poles, innovation agencies, industry, government and NGOs. The interplay with scientific activities and the openness to ideas and markets are crucial for the conception and implementation of innovations. University and industry links play an especially important role. Their interrelations can strengthen the fabric of weaker regions and create the leading edge of the stronger regions.

The think tank Intelligent Community Forum named Stockholm as the 2009 Intelligent Community of the Year. The annual conference of global best-practice communities focused on "Building the Broadband Economy: Local Growth in a Global Economic Crisis" and emphasised a "culture of use", not just access to broadband technology. Tallinn, Issy-les-Moulineaux and Eindhoven, the other European finalists, are enjoying high-technology employment in the midst of an otherwise cloudy economic climate.

During a national fiscal crisis, in the early 1990s, the City of Stockholm decided to pursue an unusual model in telecommunications. The city-owned company Stokab started building a fibre-optic network throughout the municipality in 1994, as a level playing field for all operators. The company installed conduit and ran fibre and offered dark fibre capacity to carriers for less than it would cost them to install it themselves. The network has more than 90 operators and 450 enterprises as primary customers and tries to bring fibres to 100% of public housing, which is expected to add 95,000 households to the network. The target is to provide fibre connections to 90% of all households by 2012.

In 2007, Stockholm published Vision 2030, identifying the key characteristics the city wishes to have by that year. According to the 2030 vision, Stockholm would be a world-class metropolis offering a rich urban living experience, the heart of an internationally competitive innovative region and a place where citizens can afford and enjoy a broad range of high-quality social services and a healthy environment.

During the 2009 European Year for Creativity and Innovation, the meetings on creativity and innovation in European cities 20 years after the fall of the Berlin Wall, organised by the association "Banlieues d' Europe" tried to introduce a reflection on alternative socioeconomic and cultural models. Bringing together artists, citizens, administrators, elected representatives and urban developers, the network insisted on the enabling role of cities and culture and the importance for communities to express themselves, and find the keys to understand change, reconstruct the social fabric and reinvest in citizenship.

7.2   Knowledge Cities and Helices: Investing in Innovation

Science cities incorporate science in all their functions and promote public engagement in science, also in a festive way. Genoa initiated in 2003 a science festival recognised as one of the most outstanding events of this kind in Europe. Making science accessible to everyone is the essence of the festival open to all and everyone through educational and art exhibitions, laboratories, exhibits, conferences and lectures, round tables and multimedia activities. The second event took place in 2004 within the "Genoa: Cultural Capital of Europe 2004" and included initiatives specifically designed to address schools. Apart from the hands-on exhibitions and workshops, schools had the opportunity to present their projects, adopt an exhibition or a lecture and train their teachers. Each subsequent event made its own unique contribution in attracting the public to the world of science and technology in an enjoyable way.

The UK Government promoted science cities as a valuable model for investing more and better in science and technology. Newcastle, York, Bristol, Birmingham, Manchester and Nottingham have been designated to transform the best of British ideas and inventions into new products and services. Their key aim is making science, technology and innovation a prime instrument in an increasingly competitive global world. Developing science cities requires a range of complementary policies to address the specific needs of research and development, support university–business collaboration and influence a wider spectrum of factors that contribute to the growth of knowledge-intensive industries, such as skills, finance and infrastructure. By bringing these components together in a concentrated space, science cities can attract a critical mass of innovative businesses and become drivers of growth.

Although in different stages of development, all science cities strive to enable excellence to flourish. Science City York is a partnership between the City of York, the University of York and private industry. Created in 1998, it has capitalised on the international research strengths of the University of York and the city. The Science City York model has achieved high levels of business engagement to foster an environment in which creative science and technology excellence can thrive. In 2005, Science City York had a major track record of success with more than 240 knowledge-based organisations already creating more than 2,600 jobs in York and 60 companies in its first 7 years. Its future vision, supported by Yorkshire Forward, is to create an additional 15,000 technology-based jobs by 2021.

Innovation funding may be highly localised. Science City Manchester is led by Manchester Knowledge Capital, an established partnership bringing together expertise and resources from the public, private, academic and health professional sectors. The science city is founded on four unique strengths: Manchester's ability to deliver, its sheer scale, its creative, partnership-based openness, and its focus on inclusion and social benefit. Manchester has the combination of science-based assets, higher-education research and industry expertise to deliver significant levels of sustained economic growth. With 800 knowledge-based businesses in the core alone, employing more people than the entire UK biotechnology sector, Manchester is building the success of the science city on strong foundations.

Newcastle Science City is led by a major consortium comprising Newcastle City Council, One North East, the University of Newcastle, the Centre for Life and businesses. Newcastle Science City is trying to enhance the city centre as a focal point for concentrated science-based activities, contributing significantly to economic prosperity and growth. This will be generated through research and education, applied science, supporting services and extensive and intensive interactions. Key scientific fields include ageing and health, stem cells and tissue regeneration, energy and molecular engineering. Newcastle Science City builds upon major investment by the partnership, and will involve the development of a very large scale central facility for science and business interaction.

Nottingham is a renowned world centre of excellence for biomedicine and aims to be at the forefront of the nanotechnology revolution. Nottingham Science City is a partnership driven by the City of Nottingham and embraced by the East Midlands Development Agency, Innovation East Midlands, the University of Nottingham, Nottingham Trent University and Nottingham Development Enterprise. The designation of Nottingham as a science city recognised the outstanding achievements, reputation and excellence in scientific education and research and technology by the city's two universities, the University of Nottingham and Nottingham Trent University.

Science City Bristol has a strong and diverse science and technology basis thanks to the universities of Bristol, Bath and West England. It hosts leading private sector R&D performers in the aerospace, information technology, creative digital media and silicon design sectors, as well as publicly funded organisations such as the Defence Procurement Agency and the National Blood Transfusion Service. It aims to capitalise on this and promote science and technology as a key driver of economic development. Birmingham Science City builds on existing strengths, including the city-region's national and international links, the industrial and commercial networks, the skills, culture and heritage and its six universities and other higher-education institutions.

Science cities are in the vanguard of the campaign to make science, technology and innovation the engine of sustainable growth. A case study highlighting York's achievements demonstrates how the city has created an inspirational model for a successful science city. City leaders took bold and visionary steps to create a shared strategic vision capitalising on the University of York's world class research base and benefiting the local community. The research power of the university has played a pivotal role in establishing York as a centre of scientific innovation. The university is at the heart of Science City York's mission to establish the UK as a global leader in science, education and skills by capitalising on the strengths of world-class universities.

Technology has been instrumental in Yorkshire's economic strategy, following the decline of traditional industries. York reflects this transition from long-standing mainstays of the city's economy, such as chocolate manufacturing and railways, to Science City York. The growth of technology jobs in York has reached 10% of the working population, equivalent to that of the city's tourism industry. However, Science City York still has much to do to move Yorkshire up the innovation league table.

In the US, the two leading high-technology regions, Route 128 and Silicon Valley, constitute prototypal science cities. A triple-helix model of university–industry–government relationships and science-based growth has emerged. Such a model can be conceptualised as a cycle of knowledge, consensus and innovation. The contemporary Silicon Valley ecosystem evolved into a planetary system with strong gravitational fields, pulling promising start-ups and niche organisations as planets and satellites. The shadow of the Silicon Valley sun is long enough to stretch all the way across the continent, to the other leading high-technology region in Boston.

The years before the 2008 crisis were years of boom for Boston as the global centre of high technology. Office vacancy rates hovered around 4%, and less than 1% for Cambridge, home of many high-technology start-ups. This boom, different from the "Massachusetts Miracle" of the 1980s, was driven by software, telecommunications, medical technology and financial services. Having suffered a sharp decline in the early 1990s, Boston's technology companies had found ways to complement rather than compete with Silicon Valley.

However, Silicon Valley remains the leader and undisputed champion in nurturing technology start-ups into big companies whose products and business strategies are shaping the world. The failure of computer companies in the Boston area to tap into the PC revolution early on, coupled with the recession of 1990–1991, left the area's technology sector gasping for breath while Silicon Valley gathered momentum. Economists, venture capitalists, and technology executives suggest that, presently, Boston's strength lies in fields such as Internet software and biotechnology, fuelled by the talent flowing out of the area's seven major universities.

The key resources in the Boston area are its universities, MIT and Harvard, traditionally at the top of the Shanghai Jiao Tong University ranking, and other universities that have developed specific technology expertise, such as Boston University in photonics. MIT remains the leading institution for technology business creation. A study by MIT and the Bank of Boston, the first national review of the economic impact of a research university, found that MIT graduates and faculty had founded 4,000 companies worldwide. If the companies founded by MIT alumni formed an independent nation, it would be the 24th-largest economy in the world. About 150 new MIT-related companies are founded each year. Silicon Valley seems to be a leading destination for MIT-bred entrepreneurs. The five states with the highest numbers of MIT-related jobs include California, Massachusetts, Texas, New Jersey and Pennsylvania.

Greater Boston, with a knowledge economy whose industries are squarely in the sights of global competitors, finds itself in the vortex of change. Small businesses are developing breakthrough technologies in robotics, telecommunications and ocean observation. Researchers are inventing the next wave of medicines and building materials. And artists and immigrants are reinventing local culture. Massachusetts has the highest number of patent applications per capita and demonstrates steady growth in the number of "gazelles", Douglas Henton's term for publicly held companies that double revenues every 4 years (OECD 2006, 2007).

7.3   Cities, Epicentres of Cultural Energy

If culture is "the development of human capacities and power" (Kant), culture is the participation, the action and the expression of civilisations. Culture is neither knowledge nor erudition, it involves joy and wonder. Participation in cultural life is a universal human right, which includes the enjoyment and educational benefits of urban monuments and sites. All citizens have to find in culture their expression, values and meanings which heighten their lives. Public awareness of the values of cultural heritage is vital.

Cities have always been shaped by the ebb and flow of historical, socioeconomic and cultural events. Each city is a unique civilisation, plural and singular, with a particular identity and expression. Each one has its own body and soul, symbolic significance and power of suggestion. Cultural spaces have high existence and bequest values. Cultural heritage and activities indicate how citizens and communities transform the places they live in, how they have responded to their environment and how this is etched into the landscape. Cities as strongholds of civilisation can promote intercultural understanding if they are solidly anchored in local traditions but open to the world.

Cultural sustainability requires the integration of cultural policy objectives with socioeconomic and environmental requirements. Many cities and citizens are deeply convinced that, after education, investing in culture is the investment with the greatest and the highest long-term return. Investments range from education in matters of heritage to the reinvigoration of cultural tangible and intangible urban assets. Monuments, historic buildings and landmarks play the role of a catalyst as vectors of common memory and generate new dynamics for participation and citizenship. Functional mix and diversification may ensure the continuity of a city and its projection into the future, as a live city versus a museum city. Furthermore, the sustainable preservation of monuments and sites helps to revive traditional technologies and crafts, create new services and generate employment.

Heritage is priceless and irreplaceable. National and international declarations stress the role played by monuments and sites in the cultural identity. Monuments of local, national and international importance, specific and universal, are witnesses of human civilisation. They constitute what Fernand Braudel called the artistic geology of a city. It is the responsibility of all citizens to ensure their protection. Integrated protection is only genuinely effective when it is evidence-based and wholeheartedly accepted and supported by citizens.

Urban biographies require detailed historical inventories, intelligent and imaginative projects and illuminating public dialogues. The living past of many cities is in danger as many private projects ignoring the collective culture are being carried out almost everywhere. Short-termism, greed and corruption, but also indifference from the community, have irremediably scarred the face of many cities (Mc Donald 1989). Powerful resistance by citizens, especially in cities with a strong democratic tradition, can be decisive for rejecting projects and preserving urban jewels.

In New York, more than 40 years after the adoption of legislation on the conservation of the cultural heritage, the Historic Landmarks Centre tries to resist land pressures and to put forward buildings related to various stages of its creation. The destruction of the Penn Station was decisive for collective awareness. Since 1967, 1,110 buildings and 80 areas have been declared of historical value and 24,000 properties, 2% of the properties in the city, are protected. After 30 years of life, a building can be declared a monument, but a monument is also a magnolia which has been flowering since 1845 in Brooklyn.

The protection and the enhancement of European cultural heritage, covering the broad range from cultural landscapes to artworks, are among the main actions of the EU Culture 2000 programme. In 2002, the launch of the EU Prize for Cultural Heritage/ Europa Nostra Awards aimed at promoting high-standard preservation and providing lessons for exemplary rehabilitation. This unique awards scheme, jointly organised by the European Commission and Europa Nostra, the pan-European Federation for Cultural Heritage that represents over 250 NGOs and over 160 other heritage organisations in 45 countries across Europe, celebrates outstanding cultural heritage initiatives and highlights exceptional restoration and conservation, research and education achievements, as well as dedicated service to heritage conservation.

The award aims to promote, through the power of the example, high standards and high-quality skills in conservation practice and to stimulate transborder exchanges throughout Europe. Over the years, the awards programme has seen remarkable examples of cultural preservation of Europe's heritage that represent both some of the most well-known sites in Europe and others that are unique and hidden treasures of the communities. In 2008, the winning entries included the famous Segovia Mint (1583), a unique treasure of European industrial heritage, "the world's oldest, most advanced manufacturing plant still standing today". The Royal Mill Mint in Segovia was the most technologically advanced Spanish mint until 1700, when modern screw presses were installed at the mints in Seville, which eventually culminated in the closure of, among other mints, the Old Segovia Mint in 1730.

Historic and monumental sites can be precious resources for eco-tourism, a human activity with economic, social, political, cultural and educational significance. As natural sites, cultural sites could be enhanced by sustainable tourism which includes protection and preservation of the monuments. It is of utmost importance to prevent any irreversible impact on monuments, especially due to pressures from mass tourism on archaeological sites and monuments. The interaction can be made beneficial if it is well planned and managed. Cultural tourism twined with eco-tourism can provide new opportunities for linking conservation to sustainable development and for creating spaces of harmonious coexistence between visitors and inhabitants.

Cultural eco-tourism tries to reconcile exceptional natural and cultural sites and bridge conservation and tourism. It enhances the quality of leisure by providing a resource which gives scope for individual and cultural fulfilment. Public use and physical development of heritage sites should be limited to those activities and facilities that are necessary and appropriate to fulfil visitor understanding and enjoyment of the sites without detriment to the cultural resource. With increasing visitor use, care should be taken to ensure that human pressure does not reinforce the erosion caused by nature.

Sustainable cultural tourism has to enhance the potential of cultural monuments and sites both for cultural development and as an eco-tourism resource. It can be decisive for the enrichment of the tourist patterns and the socioeconomic development of regions. The rediscovery of neglected sites could serve to redirect tourist waves. Local projects founded on the revitalisation of ancient spaces could lead to the upgrading of tourism standards. Preservation for tourism, even in small projects, requires long-term policies, given the importance, scope, diversity and complexity of monuments and sites, in their developing surroundings and changing material and immaterial context.

The sustainability of cities rich in cultural heritage depends much on their meaningful and conscious integration into modern life. Venice, the most emblematic European city, offers the epitome of conflicts between exceptional physical and cultural heritage and modern life and provides lessons about the carrying capacity of places which should always be respected. The city's environmental equilibrium has been severely threatened by industrialisation, urbanisation and agriculture.

Venice is a vulnerable environment at the interface between the Adriatic sea and the alluvial basin extending from the Alps to the coast. The Venetian lagoon is a very complex hydrological system and includes streams and their connections, ancient canals and beds of water courses. Centuries of human activities have shaped a unique yet fragile ecosystem and altered the form and character of the natural morphology of the lagoon. Progressive degradation of the quality of the lagoon water and the biotic system is mainly due to urban, industrial, agricultural and other discharges, including port activities.

Since the beginning of the century, the difference in the level between the land surface and the sea surface has changed by 23 cm, an 11-cm rise of the sea level and 12-cm shrinking of the land (EEA 2007a). This makes Venice exposed to increasing risks of flooding. At the end of the 20th century, flooding occured around 40 times a year, which is 6 times the average at the beginning of the century. The year 1966 was marked in the history of Venice as the year of the *acqua grande* when floods demonstrated the fragility of the urban ecosystem. Since then, the city has suffered a demographic haemorrhage, with a dramatic fall in the inner-city population. Concurrently, there has been growth in tourism activities, and in the number of jobs, whereas the declining population of school-age children has been counterbalanced by the increased numbers of university students.

The unique natural and cultural environment makes Venice a compelling destination for tourists, but local inhabitants feel that the overall impact on their city is negative. Much debate has gone into the idea of imposing fees on tourists visiting Venice. However, many politicians object fiercely, suggesting that Venice would run the risk of becoming another Disneyland in Europe. An integrated "access reform", offering more than the admission (e.g. unlimited use of *vaporetti* or entry to museums) could lead to a more acceptable option.

Cultural tourists, staying a few days in Venice, represent just a small percentage of the population that enters Venice every day. The main invaders of Venice are the more than eight million commuters per year and as many excursionists (1-day trippers). Tourist pressure has crowded out the local population and many economic activities. This caused urban decline and radical transformations in the functions of the city dominated by the tourism industry. The upkeep of the buildings and houses is minimal and in many cases virtually at a standstill, which adds risk factors to the environmental deterioration and the perception of Venice (EEA 2005, 2007a).

In Fredrikstad, Norway, the seventieth-century old town and fortifications are considered a legacy of the past. They have not been demolished in the name of progress but have been preserved for prosperity. Its rich history has not been an obstacle for Fredrikstad to be at the forefront of environmental action and global engagement. The cooperation with the twin city of San Martin in Guatemala has given impulses to new international initiatives.

Cultural parks and itineraries are two components of cultural tourism that involve monumental spaces and could have a particular significance for sustainability. The design of cultural parks and itineraries has to consider the quality and aesthetic characteristics of cultural and environmental resources and their integration into the urban environment.

The creation of cultural parks and itineraries has to be completed by a sustainable mobility and accessibility plan and study of additional facilities. The management and financial planning has to include measures for positive interactions between the conservation works and the enjoyment of the space by visitors and citizens. Many of these steps have been followed for the cultural parks in cities such as Athens and Rome, whose green and cultural spaces are being given special attention in their sustainability plans.

The threats to cultural heritage do not only come from tourism, local consumption lifestyles may often be detrimental. The mobility study in the historic centre of Toledo pinpointed even the cars parked in public spaces as a destructive element for the urban heritage. Once parking had been prohibited, exceptional heritage places regained their splendour. Many historic city centres witnessed a parallel experience when access to them by car was prohibited.

Cultural itineraries have their origin in the pilgrims' route to Santiago de Compostela. The Council of Europe encouraged the establishment of transnational cultural itineraries. At the level of the city, interesting cultural itineraries relate to prominent historic and cultural features. Each 16 June, Dubliners and visitors celebrate Bloomsday, the day on which James Joyce had his fictional character Leopold Bloom criss-crossing Dublin in *Ulysses*. The itinerary started in 1954, when some poets and citizens decided to mark the 50th anniversary of Bloomsday by a pilgrimage to the Martello Tower in Sandycove, where *Ulysses* starts its journey throughout the city. The centenary of Bloomsday in 2004 was marked by a Dublin-wide celebration of epic proportions.

Creating cultural itineraries on a common theme could create urban–regional partnerships and help achieve regional policy objectives by enhancing historical and cultural heritage. Such itineraries demand measures for the restoration and preservation of monuments, the creation of museums, plans for ecological resorts and environmentally friendly transport and measures for revitalising craft industries. Partnerships with chambers of commerce are important to gain the support of businesses, promote the itineraries and communicate eco-cultural messages to the public.

Each city should be enjoyed as an invaluable compelling heritage and a masterpiece of art (Olsen 1987). Art always creates new possibilities to appreciate a city's soul and to listen to its heartbeats. Urban itineraries can be metronomes of desire in the quest for urban beauty. Long rejected as a sign of frivolity and elitism, the beauty of cities, made up of asymmetries, paradoxes and contradictions, is returning to the urban stage. Art can create much urban added-value. The 1% rule in Amsterdam indicates that 1% of the cost of the new buildings should be invested in art, whereas The Hague's 1998 sculpture city event provided a creative environment for its citizens.

Encouraging a cultural understanding of public spaces is critical for sharing common values and forging a shared identity. Promoting art and culture can be a joyful and purposeful means of participation and coaction. Forms that evoke memory and stimulate imagery may foster local communities. The European Capitals of Culture, the European Nights of Museums, the Nights of Heritage, the White Nights in Rome, Madrid and Brussels or the light festivals in Helsinki invite citizens and children to see under a different light the urban environments of their everyday itineraries.

Every August, the full moon nights in Greece allow archaeological places lit by the full moon to be rediscovered. Fireworks invite citizens to rediscover civic centres, churches and palaces in Italian cities. Dance days in Barcelona encourage citizens to make innovative use of their surroundings and bring an intimate, immediate quality to their everyday lives. Carnival seasons disrupt everyday life and bring local fun to its apogee. The beggar's festival in Amsterdam helps to dispel stereotypes (Landry and Bianchini 1995).

The concept "European City of Culture" (renamed "Cultural Capital of Europe" in 1999) was created in 1985, on the initiative of the Greek artist and Minister of Culture Melina Mercouri, a great believer that culture, art and creativity are no less important than technology, trade and economics. This resulted in a series of successful events promoting cultural excellence on the stage of Europe. The experience of being Cultural Capital of Europe has reshaped many cities. Their tenure as cultural capitals has been a transformational event giving self-confidence to cities delivering demanding cultural and artistic programmes.

Athens was the first Capital of Culture, in 1985. Mercouri campaigned for the return to Greece of the Parthenon's marbles, acquired in disputed circumstances at the beginning of the nineteenth century by Lord Elgin and sold to the British Museum in London. The opening of the new Acropolis Museum, in June 2009, reaffirmed the expectation that the Parthenon sculptures will grace the Parthenon hall in its original light.

The construction of the new Acropolis museum, built after the winning design of the 2004 international competition by Bernard Tschumi, allows exhibits to be seen in natural light and incorporates a number of on-site excavations, including a large urban settlement dating from Archaic to Early Christian Athens. Most remnants are viewed from above through glass panels. The upper glass gallery, in front of the Parthenon, is awaiting the restitution of the sections of the original marbles, still in the possession of the British Museum.

The Glasgow 1990 Cultural Capital of Europe was a magnificent ground-breaking event, which transformed the city. Through this event, Glasgow demonstrated that it was determined to overcome its crisis. It gave its own definition to culture, as all-encompassing, incorporating not just music, drama, theatre and visual arts, but everything that characterises it as a unique, dynamic city: architecture, urban design, engineering, entrepreneurship, education, religion and sport.

Munich stands as the cultural heartbeat of southern Germany. Within the stimulus portfolio to enable the recovery of the German economy, which, since 2008, has suffered its worse crisis since the Second World War, an important part is allocated to infrastructure projects supporting the country's cultural legacy. The Deutsches Museum, in Munich, the most popular museum, welcoming 1.4 million visitors each year, is the largest single beneficiary.

The Deutsches Museum, focusing on natural science and technology, is home to exhibits of several world firsts, such as the first car, first X-ray and first civil airliner. The budget allocated to the museum is expected to improve its infrastructure and make it more energy efficient, and funds from the private sector will contribute to its full renovation. Business sponsors, such as Siemens and Bosch, together with the University of Munich, will be able to exhibit their latest research results in the renovated museum, on subjects ranging from stem cells to nanotechnologies and gene technology.

Cultural life in Munich is rich and a source of revenue for the city. The Bavarian capital has 42 museums and collections, more than 70 galleries and around 60 theatres. The Kunstareal, the art district of the city, has recently been enriched by the creation of the Brandhorst Museum. It also represents a major tourism asset, as it is also the extension of the bowl-like BMW museum in the north of the city.

7.4   Cities as Theatres of Artistic Creation

The arts constitute the ultimate expression of the creative intelligence of humanity (Jimenez 2002). In all their forms, they invite openness to new universes. All citizens should have the opportunity to express themselves through artistic ways and/or appreciate art during their leisure time, which has to be time of freedom and creativity.

In the Greek cities, restored ancient theatres are a prime example of diachronic places serving the same art, the drama composed of the tragedy and of the comedy, for more than two millennia. The theatre, called by Nietzsche "the supreme task", is probably a most perfect and live form of art, because it combines all means of expression. "Art among arts, festival among the festivals" wrote Vassilis Rotas, while stressing the fact that in the open air, the theatre takes its most festive dimensions, in the communion and the communication.

As a religion spreads through churches and places of worship and pilgrimage, art expands through places which encourage amazement and wonder. The EU 2009 winner of the biannual Mies van der Rohe architecture prize sheds light on the Oslo Opera House, conceived by Snøhetta. According to the jury chair, this opera house… "is more than just a building; it is first an urban space, a gift to the city." A "carpet" composed of tens of thousands of individual bricks forms a ramp becoming the building's roof directly connected to the fjord in front of the opera house. As befits a building on the water, the interiors are meant to hark back to traditional Norwegian boat design, with undulating wood walls.

Every 10 years since 1977, the Skulptur Projekte has been invading Munster and changes the face of the city. In 2007, the fourth event enriched the inheritance of the 37 sculptures produced previously with 37 new works. Art emerges everywhere in the city and invites everyone to enter a new world or to look at the city with different eyes.

Beijing's Temple of Earth hosted a symbolic sculpture project in August 2009. One hundred ice sculptures of children melted in heat represented the million lives that will be lost in Asia because of water shortages caused by climate change. The artwork commissioned by Greenpeace to launch a campaign for bolder climate change measures was one of the politicoartistic events to raise awareness and sway negotiations in the lead-up to the Copenhagen COP 15.

Arts and culture can also be catalysts for large-scale urban regeneration. In Bilbao, the creation of the Guggenheim Museum was critical for the transformation of the city. The museum revolves around a broad luminous atrium. Its titanium structure undulates harmoniously and can collect each ray of sunshine so that the natural light penetrates each exhibition hall. It is a temple dedicated to modern art which gave Bilbao a new breath and the dynamic image of a city ready to grasp the opportunities of the future. The Arts Museum in Bonn reminds us that buildings have five fronts and that one must take particular care of the front turned towards the stars. Hamburg University installed five umbrellas with solar panels (sunbrellas) on a publicly visible top of the roof of a public library and this can further underline the advantages of facing the sun.

Events organised around the European culture capitals call on citizens to discover their urban heritage and to make of their cultural environment a shared project. The 1990 Glasgow Cultural Capital of Europe provided a prime model for a strategy where the arts act as a catalyst for urban regeneration and modern creation. The city became a thrilling international stage and hosted over 3,400 public events and 1,901 exhibitions, involving performers and artists from all over the world.

At the end of 2008, Liverpool's memorable year as Cultural Capital of Europe climaxed in a spectacular fireworks display from ships on the River Mersey. Liverpool's year as Cultural Capital of Europe involved more than 10,000 artists, 160,000 participants and 67,000 schoolchildren across 7,000 events at more than 1,000 venues. The sheer scale of the project, for which €450 million was invested in the run up to 2008, was reflected in the diversity of the programme. As well as the more high-brow Liverpool Biennial, the UK's largest festival of contemporary visual art, the programme included Paul McCartney's Liverpool Sound concert, the launch of the annual Tall Ships' Race and a visit from a delightful 15-m-tall spider, courtesy of the French artistic group La Machine Now.

The Cultural Capital of Europe managed to trigger a wider debate about arts funding in the UK, which is widely believed to be highly dominated by a London arts scene. To redress the balance, the Turner Prize for contemporary art and the Stirling Prize for architecture were both presented in Liverpool. The city attracted an estimated 14 million visitors in 2008, approximately 25% of whom were reported to be first-time visitors. A record number of hotel beds were occupied and the occupancy rates reached an all-time high.

The Norwegian port of Stavanger shared in 2008 the title of the cultural capital with Liverpool. Enriched by the oil industry, Stavanger submitted an ambitious programme and a will to promote all arts, from Greek ancient theatre to operas for children and ski ballets on a fjord rock base. And the torch was carried over by Vilnius and Linz.

Highlights of the Linz 2009 Cultural Capital of Europe included "Make the Impossible Possible" according to the watchwords of the Academy of the

Impossible. The events ranged from sacred dance, biographical theatre and overtone chanting, improvisation and dancing for kids, to Balinese shadow theatre and Indian dramatic arts. The whole city took an active part in the programme of the Cultural Capital of Europe. The culture neighbourhood of the month made its way through Linz over 9 months. The city invited the cultural neighbourhoods to tell their "thousand and one stories". Celebrations in "pomp and circumstance" marked the accession ceremonies as one neighbourhood took over from its predecessor. The Art Palace, a metamorphosed caravan, provided the festive atmosphere required by such momentous occasions.

Vilnius, sharing with Linz the title of the 2009 Cultural Capital of Europe, had more than half a million visitors during the first half of the year. The majority of visitors were during the opening events of the programme of the culture capital, art exhibitions, "Street Musician Day," the project "Icy Baroque" and others. During the summer, the inhabitants of Vilnius and the city guests were invited to nearly 200 culture and art events. Reviewing the first results achieved, the city considers having discovered "a golden goose" while choosing projects necessary to realise in the year of an economic crisis and attracting the special attention of participants, but also a positive evaluation by the foreign media.

Essen, selected as one of the three 2010 Cultural Capital of Europe, will certainly be a totally different type of capital of culture. The focus will be on the shared vision for the Ruhr metropolis, a unique polycentric urban cultural landscape. It has an extraordinary structural and social diversity, varied lifestyles and living environments. The architectural landscape of the Ruhr metropolis is a showcase for innovative building design and exemplifies the ingenious transformation of industrial buildings. It embodies a new understanding of natural and human-made spaces, heralded by the International Building Exhibition Emscher Park (IBA 1999).

Art can reshape urban structures and embellish landscapes that resonate with their industrial past. Ruhr 2010 wishes to stop the clock for a brief second in the history of the Ruhr, to conduct experiments, explore new ground and rekindle a zest for life. The 2010 Cultural Capital of Europe offers the opportunity to draw upon the powerful vision of extraordinary artistic intervention that is expected to fundamentally and permanently change the way that an entire region is perceived.

Through the "Re-designing the Metropolis" programme, the Cultural Capital of Europe raises awareness in rediscovering or redefining the raison d'être of cities. The 2010 Cultural Capital of Europe invited local, national and international designers, planners, architects and artists to propose never-ending possibilities such as a winding tower as an emblem of fine art, a town square electrified by 1,000 promises, a spoil tip transformed into the "Magic Mountain" and the Ruhr creeping into the "Twilight Zone".

Other cities promote particular cultural and artistic resources. Amsterdam was the 2008 World Book Capital, a status coined by UNESCO in 2001. In the timeless words of Victor Hugo "To learn to read is to light a fire; every syllable is a spark", Amsterdam kindled the flame. The city chose to focus on the theme of "Open Book" and stirred debates about freedom of expression and the power of the written word to bridge cultural divides. Celebrations started on 23 April, the World Book and Copyright Day.

As most European cities, Vienna hosts a great variety of architectural styles, from classical buildings to modern architecture. Art Nouveau left many renewed traces in Vienna. The Hundertwasserhaus, designed to counter the clinical look of modern architecture, is one of Vienna's compelling assets. In recent years, Vienna has seen numerous architecture projects completed which combine modern architectural elements with old buildings, such as the remodelling and revitalisation of the old Geometer in 2001.

Galway, considered to be Ireland's cultural heart and a popular tourist destination, is renowned for its festivals, celebrations and events. The city is home to a huge, for its size, number of dance organisations, festival companies, film organisations, musical organisations, theatre companies, visual arts groups and writers' groups. Furthermore, it has decades' worth of venues for events, from concert venues and visual arts galleries to "multipurpose" venues. Every July, Galway hosts the Galway Arts Festival, which is known for its famous Macnas Parade. Galway is well known for its "Irishness", associated with the Irish language, music, song and dancing traditions and is sometimes referred to as the "Bilingual Capital of Ireland". The university holds the UNESCO archive of spoken material for the Celtic languages.

In 2009, the 20th occasion of the Brussels festival "Couleur Café" brought together 70,000 visitors for three festivity nights. All music styles were represented: blues, funk, hip hop, afro, reggae, dancehall, salsa, rock, tec. Couleur Café is a reference festival for intercultural dialogue which brings together a remarkable diversity of artists and participants to celebrate good music, but also tolerance and understanding of otherness in a coloured and festive environment.

First created indoors, in 1990, Couleur Café has opened itself to the sky and concerts take place in the open air. The musical scenes cohabit with a market, dancing courses, an NGO village and various demonstrations and happenings. Within the framework of the 2009 European Year of Creativity and of Innovation, Couleur Café hosted the formidable itinerant exhibition "Orbis Pictus," demonstrating that playing can be a universal means of communication, without distinction of age, nationality or religion.

Art is a perpetual opening to new worlds, an inexhaustible source of wonder. Toulouse and Brussels organise word marathons and celebrate palpitating texts. The Toulouse marathon of June 2009 invited citizens to discover their literature at the same time as their city, whereas Brussels proposed the discovery of the European literary inheritance, a poetry night, fundamental art and a reading and meeting mosaic with writers in strange places. Readings and recitations of both classic and contemporary texts took place in libraries, bookshops, cultural centres, but also in hospitals, dispensaries, retirement homes and prisons. The "Breadth of Autumn" festival or "Spring of September" taking place at about 20 Toulouse sites proves that the rebirth of a city can also take place in autumn and that the creativity resources are inexhaustible.

Watercolour 8
Madrid: A City for Science, Culture and the Arts

Madrid: A city for science, culture and the arts

Chapter 8
The Urban Renaissance of the Third Millennium

This chapter highlights models and actions for urban renaissance, a major aim for cities striving for sustainable development. Land-use planning and transport are fundamental and interrelated instruments for the sustainable regeneration of cities, of their physical parts and of their extraordinary diversity. Many cities have demonstrated that the intensification and consolidation of the urban fabric can prevent uncontrolled urban sprawl. Harbour cities face special challenges and many urban waterfronts are undergoing or have undergone radical change.

Dignified public spaces can promote collective life and local democracy and bring added value to places. Symbolic and structural projects can become beacons of the urban future. Last but not least, this chapter highlights that strategic planning for urban renaissance has to address not only the three spatial dimensions but also the time dimension of cities. Time is also a scarce and most precious resource. Local time plans can enhance the capacities of cities like chronotopes and improve the quality of life.

8.1 Recreating Cities as Beacons of Civilisation

The creation of the quintessential sustainable cities of the third millennium demands revolutionary changes in concepts, strategies, tactics, realisations, conditions and interactions. Rationality and creativity, the fundamentals of science and the arts, can help to masterfully interweave all registers – scientific, economic, sociological, environmental, cultural and political – and (re)create cities, a process which can only lead to prototypes (Mega 2008).

Sustainable regeneration is a precondition for urban renaissance, which implies a profound and lasting transformation of cities initiating a new urban era and favouring the recreation of cities like *civitas*, universal spaces on local territories, increasingly functional and diversified.

More recently, London reinvented urban renaissance. The evaluation of the first 6 years of the London urban renaissance provides valuable lessons. In 1998, an Urban Task Force report set out a vision for a well-designed, compact and connected city enhancing social well-being and environmental responsibility, where people

V.P. Mega, *Sustainable Cities for the Third Millennium: The Odyssey of Urban Excellence*, DOI 10.1007/978-1-4419-6037-5_8,
© Springer Science+Business Media, LLC 2010

can lead a fulfilling life. The assessment highlighted some successes, notably a measurable change of culture in favour of cities and a densification of city centres. In 1990, only 90 people lived in the heart of Manchester, but there were 25,000 in 2004. The consolidation of the urban fabric progressed with 70% of new developments on brownfields instead of greenfield land, compared with 56% in 1997. Urban densities increased from 25 dwellings per hectare in 1997 to 40 dwellings per hectare in 2005 (Rogers and Urban Task Force 2005).

Urban renaissance is a promising concept for martyr cities, symbols of distressed urban identities or cities which suffered from urban genocide and reached a point of no return. Experts remind us that the city is the only living organism which has the capacity to renew and reinvent itself. Oriol Bohigas claimed the virtues of metastatic planning, with the injection of key projects into the urban fabric which have a powerful potential and can ignite the overall revitalisation of the urban web. Along with homeopathic planning, it may offer valuable poles of catalytic action for the renaissance of the cities.

Urban renaissance requires a dynamic balance among coevolving policy actions which have the potential to transform the entire fabric and create new prospects for the future. Copenhagen went through decades of investment in urban renaissance. The 1989 regional plan for Greater Copenhagen promoted a better city instead of a larger city. The ongoing urban renewal in Copenhagen, described as the leading recycling project in Denmark, is based on principles of quality and equality and aims at ensuring sustainable development in relation to natural, cultural and human resources.

Berlin, the resurrected European city, strives to create a new future out of an emotionally charged past. According to Wim Wenders, "history is here physically and emotionally present". Twenty years after the fall of the Berlin Wall, the urban scars seem to be healed. All ruins, potent symbols of war and arbitrary partition, have been restored. The symbol of the hollow tooth as a monument of memory is significant. The city tried to preserve its unique alternative character with its rich layers of culture and history. Parvenustadt at the turn of the nineteenth and of the twentieth century, Weltstadt during the cultural golden age of the 1920s, Germania in the period of the Third Reich, Frondstadt during the Cold War, Mauerstadt ("wall-city") from 1961 to 1989, Wiedervereinigte Stadt (the reunified city) in 1989 and capital city since 1990, Berlin became a European metropolis of the future (Berlin capital 1992).

The reunification of Germany and the rise of Berlin as the capital city created a building boom which made the city look like a gigantic construction site. The urban core has been redesigned for people and flagship projects have enriched the urban fabric. The success of packing the Reichstag by Christo, building conventional aesthetics, facilitated its ownership as a seat of the new parliament of the unified Germany. The glass cupola, which overhangs the hemicycle, places the citizen symbolically above the elected representative. Museum Island, intended once to form a Prussian Acropolis, has been restored to recreate a place of wonder. The limestone bust of Nefertiti has returned to the Neues Museum and Spree is slowly flowing back to the foreground, trying to become, once again, the lifeblood of the city.

The page of the wall-city seems definitely to have been turned over. The city creates its references for the future, endowed with majestic structures, like the Sony Centre and the DaimlerChrysler Centre, new urban landmarks. The Sony Centre, a "city in the city", is conceived as a multifunctional complex assembling all the elements of a downtown neighbourhood within a single spatial structure, a forum open to all citizens.

Since its reinstatement as the capital of Germany, Berlin has striven to become a cleaner, greener and friendlier capital metropolis. Potsdamer Platz, once the hub of social life, the busiest crossroads of Europe and later the broken heart of divided Berlin, is becoming once again a green beehive. The German capital demonstrates that urban renaissance does not only concern the physical spaces. It creates symbols for the future. The 1-km-long part of the wall exhibited in East Side Gallery may serve as a metaphor of the many mental walls that still have to be broken down (Berlin capitale 1992).

In Amsterdam, urban renaissance has been linked to the vision of a diverse and consolidated city where the ingeniousness of people can optimise land and water resources. Until 1950, Amsterdam had developed in a series of concentric rings, forming a semicircle. Each new extension was built as a belt around the previous one. The 1950 General Extension Plan added lobes like the fingers of a spread-out hand (Pistor et al. 1994). During the reign of the private car, some concentric canals have been filled in with soil to make more space available to road traffic. Sustainability actions led to the reopening of canal rings and the intensification of land use. The water environment of the city has been rediscovered. Many Dutch cities, such as Leiden, worked hard to create mixed-use and citizen-friendly environments, including pedestrian and bicycle bridges spanning canals.

Urban renaissance is also essential for many small and medium-sized cities. Siena offers an example of regenerating life in the city centre through the creation of attractive and safe living, working and public spaces. A sustainable city should inspire pride in the community. Cultural associations in Siena (Conrade) have a power parallel to that of the elected administration. The shell-shaped Piazza del Campo, the town square, which houses the Palazzo Pubblico and the Torre de Mangia, is another architectural treasure, famous for hosting the Palio horse race for hundreds of years.

In Toledo, the strategic plan offered a good diagnosis of the city's strengths and weaknesses. The city tried to achieve a vital dialogue between the historic and the contemporary city. The enhancement of the cultural heritage, consisting of historic buildings, vernacular architectural spaces and the fabric of the streets, the accessibility and the optimisation of the city's potential, the harmonious coexistence of urban functions in the historic centre and the promotion of the performing mix among the university, the administration and tourism are organic parts of the plan (EFILWC 1997c, d).

8.2 Sustainable Regeneration Rather Than Expansion

Progress towards sustainability demands greater intensity of land use and the conversion and reuse of brownfields and buildings. It also requires the preservation of surrounding greenfields and biodiversity. Cities that strive to become sustainable opt for renewal rather than expansion, for consolidation of the urban fabric and improvement of the peripheral districts and suburbs.

The paradigm of the compact city, in contrast to the diffuse city, is considered to be most compatible with a resource-conscious city (Danzig and Saaty 1973). Compact cities have the potential to mobilise huge amounts of resources at remarkably lower levels of energy consumption, resource use and waste generation compared with diffuse cities, rural settlements and dispersed populations. Furthermore, it seems that a compact city grows more compact (Pistor et al. 1994).

The degree of compactness and the overall urban density are critical indicators for sustainability. True compact settlements imply a clear definition of their boundary to discourage sprawling processes, regeneration of open and derelict spaces, functional diversification of land use at the neighbourhood level and environmental improvement of the external subcentres, well served by public transport. The Danish model of decentralised concentration highlights the importance of all these components, whereas the Dutch compact city policy promotes the principle of spatial multifunctionality.

The interrelated questions of density and compactness are of prime importance. Although often combined, high density should be distinguished from high rise. Hong Kong, the most vertical city in the world, with the greatest number of skyscrapers (7,650), is one of the most densely populated areas. This is the result of a unique endeavour to adapt to a very limited surface and enhance an exceptional topography.

Still higher buildings is one of the old dreams of humanity and, well after New York's emblematic skyscrapers, Asian countries proposed dynamic towers, with mobile floors which allow residents to change views. On the threshold of the millennium, Saudi Arabia inaugurated two of the largest towers in the world in Riyadh, which, according to the words of the architect Norman Foster, "push back the limits, conquer space, penetrate the skies in order to open new horizons". The limit was declared to be no longer money but human imagination.

Europe also has its towers spread over history. From the medieval towers of San Gimignano to today's Turning Torso in Malmö, with its extraordinary spine of steel that twists through 90°, towers represent a bold theatrical architecture regarded by locals with pride. They soar above the surrounding skyline and timeline and add a note of magnificence to the identity of the place.

Tall buildings have a higher embodied efficiency and help to create the concentration of people and uses necessary to sustain greater environmental benefits. Lower-rise and high-density combinations may also provide the most interesting results. High density may convey the illusion of chaos, but bestows the benefits of close links that guarantee social and spatial stability. "Proximity means that you drive less and care more about those living next to you" (Owen 2009).

Mixed land uses well integrated with transport are a guiding principle for many city plans (World Bank 1995). Most European cities opt for relatively low buildings. Vienna had, in early 2006, only 100 buildings higher than 40 m. The number of high-rise buildings is kept low by building legislation aimed at preserving green areas and districts designated as world cultural heritage sites. The 202 m-high Millennium Tower located in Handelskai is a symbolic building to reach the stars.

Sustainable cities must reflect a true urban and social intermixture. There is an increasing unanimity about the need for an urban mix, a desire for a little of the city everywhere in the city. Functional zoning seems an anachronism from the bygone industrial era. In many cases, it created more problems than the ones that it was supposed to resolve and often destroyed the complex cultural, economic and democratic fabric of a city. The reintegration of urban functions should reinforce identity and openness. The concept of the "open block", theoretically developed by the architect Christian de Portzamparc in the 1980s, advocates harmonious compact neighbourhoods open to the city (De Portzamparc 1996a, b, 2007; De Portzamparc and Sollers 2003).

Revitalising an urban area entails recreating the economic diversification, the social heterogeneity and the cultural diversity of the city. The concept of urban villages advocates the polycentric development of a city by multiplication of its vital cells, its neighbourhoods, and not by overexpansion of its central core. Each neighbourhood should bear the genetic code of the city, as a backbone to its unique identity, and be in harmonious symbiosis with the other neighbourhoods. Every urban district should offer citizens the urban functions and services necessary for quality of life, work and leisure (Mega 2008).

Sustainable regeneration has to address the unrealised potential of distressed urban areas willing to reattract social life and productive investment in healthy environments. Successful regeneration schemes address both the hardware and the software of the areas and reconcile environmentally sound rehabilitation of physical structures with social revitalisation and economic enhancement, while opening the area to new ideas and cultural flows.

Diversity and mixed land uses are linked to the city's unrivalled character as a cultural and racial melting pot. Amsterdam considers urban mix to be a valuable attribute of the inner-city heritage and tries to strike a balance among offers for housing, offices, commerce, services, tourism and leisure. Since 1965, a pedigree sustainability principle has been followed. The new plan was based on the concept that cities should not continuously expand but should concentrate on consolidation and renewal. The compact city policy, introduced in 1985, aimed at optimising and maximising the use of space, decreasing distances between home and work, creating diversified residential environments and reducing the overall environmental burden. The implementation of this policy has been a real struggle with scarce space and can provide many valuable lessons.

Commercial poles at the periphery of cities are converted into vital neighbourhoods. The Huddinge Centrum, in the south of Stockholm, is a successful experiment in transforming a suburban shopping centre into a town square, a vibrant public space for the community. The location, next to the train station, generated the creation of new offices and apartment units and the whole area was reshaped after the model of the medieval part of Stockholm. The "Living Above the Shop" project, in Dublin, was also recorded as a prime example encouraging and assisting shop owners to convert their upper floors into residential spaces and improve the functional mix of the areas (EFILWC 1996a).

Diversified developments have been introduced even at the scale of one complex, typically comprising a well-connected shopping mall, leisure places, offices and apartments. In Galway, the main commercial centre encompasses archaeological remains of the ancient wall, the shopping malls, offices and, on the top terrace, a complete housing estate for inhabitants who wish to live in the city centre and yet enjoy their own homes with their symbolic front doors (Mega 2008).

Regeneration, revitalisation and even rejuvenation are important paths to sustainability. The common aim is to recreate attractive urban places out of anonymous and dysfunctional spaces and make cities vital places and strong magnets. Sustainable regeneration incorporates ecological principles in all stages: identifying and evaluating the place, planning for change, codeciding on a resource-friendly future, reinventing and enhancing the city. Citizens should be deeply motivated and involved and have pride in the places where they live and work.

Sustainable regeneration tries to inject new life into petrified spaces and transform idle city assets into sustainable resources and benefit generators. The Dublin regeneration programme aimed at redeveloping dilapidated core areas, halting the dramatic decline of the population in the city centre, reversing the spiral of decay and creating a climate of confidence to stimulate and win back investment. The Dublin City Development Plan, published by the municipality in 1991, after 6 years of preparation and consideration of 21,000 representations and objections, provided a valuable framework. The government reacted with major initiatives, which promoted the Celtic tiger economy, including the designation of assisted development areas, the establishment of the International Financial Services Centre in a derelict dock area and the creation of the cultural quarter of Temple Bar (EFILWC 1996b).

Brownfields and underused space represent valuable resources which have to be enhanced. The removal of architectural and derelict barriers and designation for public purposes, such as green belts, can undoubtedly be beneficial. This evolution contrasts with the one, one century ago, when many European cities, after the example of Vienna, had their walls, a then underused asset, demolished to make way for large motorways, circling city centres.

Opening up problematic areas and linking them to the city is vital for oxygenating the urban tissue. In Vienna, the rehabilitation of the Nordbahnhofgelände urban wasteland is a good example of restructuring a former military and industrial area. A competition stressed the optimum and maximum use of existing space, the free and green space, the maintenance of a biotope of quality, the wide use of solar energy and waste optimisation. Leisure spaces include a cultural centre temporarily installed in an old tram depot.

8.3   Cities on the Waterfront: Chances and Challenges

Water often oriented the way that cities grew and developed. Cities depend on the waterways, and the waterways achieve significance through the cities. Rivers are mirrors of ecological awareness and cooperation among the territories that they cross. Waterfront cities are shaped by geography, nature, climate, history, economy and culture. As a result of economic and technological reforms over the last few years, city-centre ports have disappeared, leaving behind deserted shipyards, legacies and relics of the Industrial Revolution, which became the starting point for sustainable regeneration.

The sustainable conversion of waterfront areas, seasides and riversides, is a key feature of many cities. In the context of the global conference on the urban future (Urban 21) held in Berlin in 2000, ten principles for the sustainable development of urban waterfront areas were adopted. They include securing the quality of the water and the environment, the integration of the waterfronts into the urban fabric, the importance of the historic character, the priority of mixed uses, the prerequisite of public access, the planning through public–private partnerships, the importance of citizen parti cipation and international networking, the long-term horizon and the ongoing character of the revitalisation process.

From the former Almada docklands in Lisbon to Turku, Malmö and London, disused dock infrastructures are being turned into exhibition halls, shops, craft workshops and centres for ecological, leisure, civic and cultural activities. Functional diversity is an increasingly important characteristic and public access to the waterfront is considered to be decisive for the shared ownership of the sites. Waterside promenades are increasingly replacing industrial dock spaces and welcome services and citizens. Barcelona, which has always lived with its back to the sea, provides an outstanding example of reconciliation between a city and the sea (Ajuntament de Barcelona 1995).

By the end of the 1970s, all major British cities inherited large areas with aban-doned dock industries and warehouses. The most spectacular case was the London docklands, an area of 8.5 square miles, beginning at the very edge of the London central square mile and the City. It included the relics of the once greatest port of the world. The regeneration of the London docklands led to the thorough transfor-mation of the area stretching across parts of the East End boroughs of Southwark, Tower Hamlets and Newham (Hall 1998).

The London Docklands Development Corporation, in charge of the project, has worked for 17 years to bring a new face and significance to the place. An insightful account of its work is recorded in a series of monographs mapping the issues, from the initiation of urban change, the creation of the light railway to central London, engineering, planning and development, the property market, employment opportu-nities, housing, regeneration and the arts and also the role of local communities.

Recognising waterfront heritage as a precious diminishing asset is crucial for encouraging selective preservation along with quality urban design. In Galway, the waterfront is framed by the River Corrib and a number of canals and channels built through the city. The canals provided a power source and served as the location of the first industries in the mid-nineteenth century. The renewal of abandoned urban areas surrounding the canals, after the move of the industrial activities, provides an insightful example. A dilapidated central area became subject to overall renewal, once traditional activities had disappeared or had relocated to sites outside the city. Economic measures were introduced, promoting rehabilitation as well as new constructions. The regeneration respected Galway's unique character and atmo-sphere and promoted a functional mix, essential for the vitality of the urban centre.

A balance has been struck among residential, commercial, administrative, academic, cultural and tourism functions.

Water long shaped the way Amsterdam developed and still plays an important role in the hydrological equilibrium of the city. It no longer shapes the city the way it used to, but rather is incorporated in economic and eco-cultural projects. Bilbao used to be a gloomy and rainy city endowed with an obscure river. After a radical metamorphosis, it became a luminous city with green and leisure spaces, a welcoming riverscape, a friendly new metro, a tram line and the Guggenheim Museum, which marked a watershed of excellence. Citizens have suggested that it even became sunnier.

The restructuring of the Belfast waterfront has been driven by the strong will to heal the scars of a divided city and create a new forward-looking face. Spaces of hope replaced places marked by violence. The integrated management of the dual city–harbour resource has been essential. The commercial and industrial docks dominating the Belfast Lough shoreline include the famous Harland and Wolff shipyards which propelled Belfast onto the global stage in the early twentieth century as having the largest and most productive shipyard in the world. The maritime festival, every summer, is a celebration of the links between the city and the sea (BURA 1997).

In Finland, Turku also had to face the decline of its industry along the River Aurajoki. In 1987, the municipal council organised an architectural competition for a new master plan for the area with the relics of closed-down factories and warehouses. The winning entry, Sigyn, introduced a magnificent mixture of old and new structures in brick, steel and glass, for educational, economic and cultural purposes. Two massive former shipbuilding halls and a former rope factory, once voted the ugliest building in town, composed a major fine arts complex, including a conservatory, the Turku School of Art and Communication and the School of Fine Arts.

In Sweden, Malmö has been a proud shipbuilding city, the birthplace of large vessels and provider of numerous jobs. With the decline of shipbuilding activity, the economic situation became critical. With the help of the government, the city managed to build a prosperous future. A new university has been created and the construction of the Öresund Bridge and the City Tunnel contributed to reversing the problematic urban economy. The former shipyard area of Kockum has been rejuvenated and hosts many new activities and Malmö University with its 20,000 students (Malmö 2008).

Liverpool's waterfront, a world heritage site, reflects the city's importance in the development of the world's trading system and dock technology. The well-known Albert Dock, restored in the 1980s, is a living textbook of harbour industrial architecture. Part of the old dock complex hosts the Merseyside Maritime Museum, an anchor point of the European Route of Industrial Heritage, the International Slavery Museum and the Tate Liverpool. Other relics include the Stanley Dock Tobacco Warehouse, still the world's largest brickwork building and Pier Head's "Three Graces": the Royal Liver Building, built in the early 1900s and surmounted by two bronze domes with a liver bird, the symbol of the city, on each, the Cunard Building, headquarters of the former shipping company, and the Port of Liverpool Building, home of the former Mersey Docks and Harbour Board, which regulated the docks.

Creating a new skyline to the northeast, across the River Danube, and a new mixed waterside edge quarter, has been the aim of the Dona-City project in Vienna, inspired by the model of La Défense in Paris. New edge cities have multiple purposes and wish to attract both enterprises and residents to an attractive and healthy environment.

8.4   New Landmarks for the Urban Futures

Symbolic and structural projects can offer significant hallmarks for the city and have the potential to shape urban territories. They may act as strong catalysts for the future of the cities and the regions. They usually stem from high-scale government plans or unique international events, such as Olympic Games, universal exhibitions and high-level fairs. Their planning with respect to sustainability principles is essential given their dimensions and projected life horizon.

The creation of these projects is often an enriching process which draws on international expertise and the best available experience. Examples include the redevelopment of the Tokyo Station Area and the Tokyo Waterfront, presented by the Japanese National Policy Research Council as vehicles for introducing private capital to public-sector projects. From La Défense to Villette and Bercy, the 1989 landmark projects in Paris on the anniversary of 200 years since the French Revolution resulted from a well-determined effort to realise the symbolic potential of the capital of light, to reequilibrate the urban fabric and enhance its magnificent skyline and shoreline.

Large-scale structural projects demand long-term planning, flexibility, forecasting and ex ante impact assessment, eco-design, citizen participation and communication. Flexibility is imperative for adapting high-scale projects to market fluctuations, whereas continuity is linked to a vision for the future of the infrastructures. Stable financial structures for the realisation of major projects are of the highest importance, whereas the cost of land and infrastructure is a key issue for economic equilibrium. Consultation and partnership become important on several levels, vertically and horizontally. The success of the projects greatly depends on a constant and affirmed political determination, capable of withstanding changes in elected representation (METROPOLIS 1996).

Emblematic and unique events bring special opportunities for cities, regions and nations. Creating an infrastructure for the hosting of a short-term high-impact event but planning it for other uses on the long term is the main challenge. The 1992 Olympic Games, the first Olympic Games after the end of the Cold War, was a key catalyst for the transformation of Barcelona. Following the opening of the city towards the sea and the creation of the Villa Olimpica, the whole urban fabric has been enriched through the injection of key improvements.

The rehabilitation of the Ciutat Vella, comprising four historic quarters in the city centre, was an unprecedented event, in terms of dimension, time and civic spirit. Selective renovation, rehabilitation, construction, pedestrian precincts, civic centres and green public spaces are the visible elements. A great number of the Olympic installations are highly specialised and Barcelona gave greater attention to secondary installations for citizens and children. The invisible elements that made everything happen are the strong neighbourhood groups that cooperated with the authorities and played a pioneering role in the allocation of new housing and services and the improvement of everyday life in the city centre (Ajuntament de Barcelona 1995).

And in the new horizon of Barcelona, other innovative projects came to create a new dynamic. The most controversial is probably the creation of the district @22 in place of the former industrial zone known as "Catalan Manchester". One hundred and fifteen building blocks have been regenerated and coexist with skyscrapers and bioclimatic installations and the multipurpose exhibition centre Forum (Rowe 2006). Such projects projected the image of Barcelona to the world. The 2004 Universal Forum of Cultures, a 5-month urban event in Barcelona, honoured urban innovation and hosted a dazzling exhibition of best practices and ideas and UN-HABITAT's World Urban Forum.

In Beijing, on the occasion of the 2008 Olympic Games, more than 30 Olympic installations were created taking into account the post-Olympic function and use. The 1992 World Exhibition gave Seville and the island of Cartuja the opportunity to become a dreamland and a symbol for urban innovation. Seville was itself on exhibition during the Expo, as the mirror of a multicultural past, a magnifying glass for the present and a telescope for the future. A thematic park opened just 8 months after the closure of the exhibition and has rapidly attracted many visitors. A technology and business park occupies the rest of the space used for the exhibition.

Lisbon grasped the opportunity offered by the 1998 World Exhibition on "The Oceans" to redevelop a significant stretch of the waterfront chosen as the location for the Expo. A marginal derelict urban area, which had played a role in the past life of the city but went into decline as its activities became obsolete, was transformed into a site for innovation and modern creation. The project was not confined to the exhibition precinct of 50 ha but aimed at creating a whole new resourceful city of 330 ha. Bioclimatic conditions were enhanced to the maximum and advanced energy management concepts were implemented from the initial stage. An eco-efficient distribution network for thermal energy, heat and cold air was set up, together with an observation and monitoring system. The standards adopted were far above those required by the national regulations (EC 1999).

The 2000 Olympic Games endowed Sydney with a sustainable new organic part. Sydney Olympic Park, a 640-ha site located at Homebush Bay, previously intended for an urban renewal project, was reshaped for the Olympic Games and continues to host 1,800 sporting and cultural events per year. Before its transformation, Sydney Olympic Park was an industrial wasteland, after 100 years of industrial and military land uses. Since the end of the Olympic Games, Sydney Olympic Park has been converted into a multipurpose facility. The Sydney Olympic Park Master Plan encourages a broad range of commercial, residential, recreational, leisure and cultural uses that create new value and add to the unique qualities of the Sydney Olympic Park for residents, workers and visitors. Urban design and landscaping principles adopted in the master plan emphasise design excellence and the creation of an active centre around the railway centre.

The 2004 Olympic Games endowed Athens with an integrated Olympic Public Transport System (metro, tramway, suburban rail, light rail, buses), after the radical restructuring of the transport network throughout the Greater Athens region. The legacy of the 2004 Olympic Games to the Greek capital includes 120 km of new roads, 90 km of upgraded roads, a 40-km suburban railway (reaching the new international airport), 40 flyovers, 7.7 km of new metro lines, a 23.7-km tram network, modern train stations and a new state-of-the-art traffic management centre. The Olympic village has been transformed into high-quality homes for low-income working families, which were selected by a draw.

No urban plan can have a future if it does not pay attention to public places, the agoras of the sustainable city. The agora, the focus of civic life in the ancient Greek polis (city state), constitutes a powerful archetype for the central public space. The polis has been a complex and fascinating universe, recognised as the single greatest contribution of ancient Greece to the world civilisation. It is reputed for its form, adapted to the natural landscape and the human scale, and its noble public spaces, bastions of collective life and democracy. Harmony, proximity, mix, safety and citizenship were inherent in those city states, where the assembly, the theatre, the stadium and the sanctuaries had the noble aim of promoting the physical and mental well-being of citizens.

Public spaces belong, by definition, equally to all citizens. They are the places where people come together to celebrate and to protest, to declare or to pray in silence, to express joy or indignation. Rem Koolhas describes them as fortresses of freedom. Noble, safe, attractive and enjoyable public spaces, promoting collective life, bring a higher added value to places. They have great potential as islands of civilisation in the archipelago of the city. Open spaces can facilitate the flow of energy throughout the city and promote interactions and synergies.

Attractive public spaces may foster citizen participation and democracy. Agora and citizenship are intrinsically linked. Public space is highly charged, with multiple risks of conflicting interests. Open space is often considered to be an "empty" space with a high opportunity cost. It should not be defined as the space left after the construction of the buildings but should be given major importance as a civic space to be shaped as a matter of priority. Open public spaces should be designed with special care to facilitate and promote access for all citizens and become a space for negotiating democracy, instead of an arena of confrontation and of exclusion.

Europe is blessed with many urban public spaces which invite one to go on marvellous journeys in space and time. Athens and Rome, cradles of European civilisation, followed parallel processes for reorganising their millennial heritage into cultural parks. In Rome, the oldest and most important structures of the ancient city are located in the Forum Magnum. The forum served as a city square and central hub where the people of Rome gathered for justice and to express their faith. The forum, also the economic hub of the city and the political centre of the Republic and Empire, and the monuments surrounding the Acropolis, were enhanced to become the focal points of urban archaeological itineraries, the natural and cultural epicentres of the two capital cities.

Prague is a living textbook of 1,000 years of European architecture. From the Charles Bridge, the seed from which the city spread, most public places are linked to historic events. The proclamation of the 1948 republic is linked to the Old Town square, whereas Wenceslas Square is considered the birthplace of the 1989 Velvet Revolution.

In Siena, the medieval heritage became the key resource for a sustainable vision. The Piazza del Campo, regarded as one of the most beautiful civic spaces in Europe, grew in importance as the centre of secular life. The place also served as the site of the market and the location of various sporting events. New streets were built to facilitate access for everybody to this centre of urban life. The wall constructed in the Palazzo Pubblico in the twelfth century to stop soil erosion indicates the importance of this civic space. This does not consist solely of the built capital, but also of the traditional Palio, the annual horse race in the Piazza del Campo.

The cities of the third millennium often cannot have a single agora, they need multiple centres to accommodate heterogeneity and diversity. Polycentric cities may have multiple agoras and offer access to public buildings, services and activities. Open spaces have an important socioeconomic function, as they render a city visible and allow people to move. Their environmental and cultural landscaping is also very important in forging identity. Qualitative recommendations for the functional and aesthetic character of squares, roads and pavements, green spaces and watercourses, roadside plantations and public lighting have been developed and implemented in many cities. The Manual of Public Spaces in Brussels sheds light on an unknown urban patrimony starting from the winding streets inherited from the medieval period and the squares, hearts of the districts (Région de Bruxelles-Capitale 1995).

Stimulating the flow of urban energy sometimes necessitates little intervention. Symbolic cases such as the example of the Virreina Palace in the Ramblas of Barcelona, used for administrative purposes, may inspire. One of the two doors of the building, the back one, was closed for years, creating a barrier, a cul-de-sac. A simple key has not only transformed the character of the building, but has made a highly appreciated public passage between the Ramblas and the Boquería. Sometimes keys lock imagination and energies which can be released without any cost and are conducive to optimal benefit.

8.5   Strategic (Spatial and Time) Planning

Strategic planning is an important instrument to achieve sustainability goals. Strategic planning concerns the future of a place and not just the location of activities and infrastructures. It encapsulates visions for the longer-term development, establishes objectives and prioritises actions towards the implementation of policies. Strategic plans have to reconcile thematic and territorial policy objectives and be discussed, codecided, implemented and coevaluated. The Copenhagen "Five Finger Plan" can be quoted as a pedigree example of sustainability plans which try to develop not only a capital city, but also promote the harmonious development of the city and the surrounding region. Preserving the green wedges, consolidating the fingers and equipping them with highly performing public transport are essential for sustainability.

Urban planning for sustainability has to break with old models. A clear shift from the planning principles which dominated the second half of the twentieth century has been witnessed. The 1933 Charter of Athens and Le Corbuzier's vision of the radiant city introduced functionalistic planning, demanding the segregation of spaces for work, living, leisure and communication. The creation of Brasilia, as a modern sanitised capital city, is the epitome of that line of thought, which has completed its circle (Le Corbusier 1971; Niemeyer 1997).

The new Charter of Athens, issued by the European Council of Town Planners in 1998, places emphasis on mixed urban functions and inclusive human settlements for all, based on true involvement. The 1998 Charter of Athens advocates planning which promotes socioeconomic progress and environmental enhancement and safeguards traditional elements and identity in a city. Mobility and access, diversity and mix, health and safety are deemed essential for sustainable cities (ECTP 1998).

The design of urban policies should be preceded by resource and energy flow analyses and followed by continuous impact monitoring and reporting. Strategic planning generally entails a substantial research programme and wide-ranging consultation processes. It has to elicit fundamental principles providing the bedrock for the future, to collect and analyse data to support enhancement of the resources and to ensure political endorsement from community and stakeholders. The open and transparent negotiation may delay plans, but can also nurture a climate for permanent improvement and self-generated change. Continuous stocktaking of progress and adjustment of components should allow the achievement of objectives, especially in the face of the crisis and strained budgets.

Urban planning is of utmost importance for improving the urban metabolism and reducing the ecological footprint of cities. Urban infrastructures are long-lasting and influence resource consumption for decades to come. Public decisions on the future of urban infrastructures can make or break a city's prospects for the future. Urban planning can create and maintain future resource traps or opportunities for resource-efficient lifestyles.

"Planning our urban future" was the theme chosen by the UN for World Habitat Day 2009. The UN chose the theme to raise awareness of urban planning to deal with the major challenges of the century. Most urban planning systems are inadequately equipped to deal with the challenges of climate change, biodiversity, food security, population growth and economic crises and, to a large extent, have failed to acknowledge the need to meaningfully involve all stakeholders in the planning of urban areas. In several parts of the world, planning systems have rather contributed to the problems of marginalisation and exclusion in rapidly growing poor and informal cities. Improving and replacing old infrastructures with environmentally friendly ones and adopting better lifestyles could decrease radically urban consumption.

Transport and land-use planning constitute two strategic and interrelated instruments for sustainability. Lack of integration of the two instruments gave rise to suburban garden cities or satellite new towns, totally dependent on private cars, which have been transformed into dormitory suburbs without any collective social and civic life. Consolidation, i.e. concentration and intensified use of space and improvement of a city in a well-defined urban territory, provides advantages for the integration of urban structures.

Sustainable development requires urban structures and patterns that minimise consumption rates and transport flows per person and lead to a drastic reduction of greenhouse gas emissions and depletion of non-renewable resources, maintenance of biodiversity and enhancement of local materials and labour skills.

Urban sustainability can only be achieved through the coexistence of people, opportunities and activities. This may not be sufficient, but is certainly a prerequisite for constructive interactions. A functional mix does not only minimise distances that individuals need to cover, but also fosters proximity and increases the efficiency of the urban systems.

Local governments, while recognising constraints, should not accept that the development potential of a city is fully predetermined by its physical form or other existing conditions. They should strengthen the capacity of citizens to transform and reinvent their cities. The form of a city should be able to reflect the aspirations of its citizens. Urban eco-design should reconcile diverging interests and promote the concept of social space which most fits the essence of a city as a human and sociocultural ecosystem.

Sustainability design, the graphic language and an integrated part of urban planning, can enhance the balance among the socioeconomic and environmental dimensions and reveal the aesthetic qualities of cities. Strong leaders and enlightened planners and private developers should respect the city's genetic code, enhance continuity and change and understand and realise the potential of physical and cultural assets.

Alicante implemented a comprehensive series of urban renaissance actions. The pivotal projects were the rehabilitation of the historic heart, the renewal of the waterfront and the development of a new urban quarter on a troubled residential area, the Barrio de Mil Viviendas. The two first projects, successfully accomplished, have transformed the city. The Barrio de Mil Viviendas project aimed at reconverting a degraded neighbourhood into a vital community through the creation of self-regulated, personal and proactive housing. The local authority had designed an ambitious plan, with the participation of the inhabitants and engaged unemployed residents in the reconstruction of the quarter.

Modern, intelligent, open and accessible-to-all spaces, locally managed, were about to replace the old rigid structures when, following a change following local municipal elections, the project was interrupted. In presenting the unfinished project in a conference, the former mayor recognised that if such a project is stopped, without overwhelming public protest, the mistake lies with the authority which initiated the project without the full conviction and commitment of the inhabitants. This highlights the importance of community support as a prerequisite for undertaking ambitious innovations (EFILWC 1996a).

The urban renaissance of the third millennium will need to consider the time dimension of cities. Cities are chronotopes, with interconnected spatial and temporal dimensions and interrelated historical and geographical aspects. Time is also a scarce resource for cities. The time arrow introduces concerns about intergeneration distribution of capital and serves as a litmus test for the well-being of individuals and societies. It also has many gender dimensions, as women have a different relationship to time. Time management has a potential for extending the limits of spatial planning. Diachronic and synchronic territorial dimensions are starting to be taken into account in strategic policy-making. Some governments and cities have been pioneers in promoting local time plans strengthening places through the harmonisation of time budgets (INU-Politecnico di Milano 1997).

All forms of time, including geographical movement, social and individual time, coexist in the city. Time plans in Italian cities, such as Milan, Florence and Bolzano, try to optimise public services offered by cities to citizens. They have been linked to mobility plans and led to the modification of timetables of municipal services. Women associations have been very active in changing the time of cities and bringing them into harmony with the times of life (Zayczyk 2000).

In Rome, also a laboratory for closing certain parts of the metropolis to cars on certain days per week, the municipality, the trade unions, the city's time office and the office for citizen's rights signed an agreement on the reform of the municipal time schedules in 1995. A period of experimentation with working time schedules in municipal sectors initiated better services to citizens. The opening hours of all municipal offices have increased considerably.

Urban time management may lead to the vanishing city or the virtual city, in which technology dominates space, but also the ubiquitous city. The intelligent use of new technologies may be a source of individual and collective benefit and fulfilment. The *immotique* and the *domotique* offer new, endless possibilities. Teleworking can lead to a disassociation between concentration in time and concentration in space. Electronic commerce and home banking could account for one third of all banking transactions (Chesneaux 1996).

Teleactivities are just instruments, conducive either to integration or to exclusion, depending on the overall policy articulation and the metapolicy framework. Physical and digital infrastructures can be very complementary and link harmoniously to generate a better quality of life and work.

However, it seems that nothing can replace face-to-face contact, the eternal creator of trust and relationships. Teleactivities will never replace the real place activities. After having extended the limits of place, many cities have tried to extend the limits of time and have deployed efforts towards achieving the 24-h city and the nightlife economy, probably inspired by North American patterns and Asian markets.

Watercolour 9
Malaga: Urban Renaissance for Spaces and Functions

Malaga: Urban renaissance of spaces and functions

Chapter 9
The Cities of the Citizens

Citizens are the political stakeholders of cities. They have rights and duties to the city. Since the time of the Athenian city state, the permanence of cities in time has been related to their capacity to promote open democracies. Active citizenship means participation in and responsibility for decisions on the future of a city.

This chapter examines the emergence of new models of governance and citizen participation in accountable cities, a sine qua non condition for sustainability. Innovative partnerships can maximise the potential of synergies, enrich content and methods of cooperation and serve as catalysts of change. Institutional alliances are enriched with a variety of participatory schemes. World coalitions of cities and citizens, both from the developing world and from the developed world, can play a major role in addressing global common challenges and achieving the millennium goals for the renaissance of the planet.

9.1 "What is the City but the People?" (William Shakespeare, *Coriolanus*, 3.1.199)

Every city at any moment of its trajectory is composed of people being together to enhance their opportunities. The progressive concentration of people and activities acts as a magnet for socioeconomic development. The individuals attracted cannot, however, be perceived as an undifferentiated mass. Their distinctive characteristics have to be well recognised, especially in increasingly diverse and multiethnic societies. Decision-makers should invest in a better understanding of the identity, needs and preferences of citizens.

Citizens identify, first and foremost, with their cities. Proximity to other citizens and local authorities is the prime reason why many citizens are more directly engaged with their city than with their nation. The rise of the creative class, an urban phenomenon suggested by many thinkers, generated a multitude of creative interactions (Florida 2004, 2005, 2008). Openness and accessibility are key issues. In many cities, meeting the mayor or at least elected officials, without much difficulty,

V.P. Mega, *Sustainable Cities for the Third Millennium: The Odyssey*
of Urban Excellence, DOI 10.1007/978-1-4419-6037-5_9,
© Springer Science+Business Media, LLC 2010

means that cities get feedback in near real time. Local policies may affect more directly citizens' lives and are often the subject of passionate debates.

One of the lessons of the Greek civilisation is that the population of a city has to be regarded as a total of unique individuals. In ancient Athens, during the golden age, long before acid rain affected the Parthenon marbles, true citizenship meant being an active member of a city. In its "Epitaphios", the famous funeral oration and epitome of the Athenian national consciousness, Pericles urged citizens to become lovers of the city (Thucydides).

Citizens are the political stakeholders of cities. They have the right to information and scrutiny and the duty to exercise the democratic control of local policies. Users of public infrastructure and services may contribute decisively in creating a collective momentum for sustainable development. Policy-makers should invest in educating citizens on all important aspects of sustainable development.

A sustainable city should allow as many voices as possible to be heard and as many values as possible to be represented. Urban democracy, representative and direct, is a key element of the permanence of cities and their capacity for sustainability. Democracy can increase considerably the collective credit of a city.

An interesting link between democracy and sustainability may be found in an old Greek custom. Before becoming a fully fledged citizen, every young man had to solemnly promise to leave the city richer and better than when he first became part of it.

Social and economic actors have great responsibility for the future of a city. Their action is essential in stimulating the local dynamism before formulating any option, identifying and adapting actions and sharing the costs. Business has a catalytic role in promoting urban investments which ensure long-term green growth and sustainability. Trade unions seem to be particularly keen on the quality of urban services offered to the citizen, the workers and the unemployed. The social partners can coinitiate innovations, making them grow, finance unprecedented activities, create the culture of change. The sustainable regeneration of wastelands, depressed areas and the creation of technology and business parks are the most outstanding examples of turning liabilities into assets in partnership with industry. Trade unions have been particularly interested in employment generation and social housing schemes.

Voluntary approaches with the active participation of industry are most important. In Japan, thousands of voluntary schemes have been concluded between local communities and their businesses, where the latter have agreed to achieve specific targets, often with the local government acting as mediator and guarantor. The Rhine contract, between the municipality of Rotterdam and polluting industries on the Rhine is also a prime example.

Volunteering is an outstanding form of expressing active citizenship. It is estimated to have an underexploited potential. Voluntary activities are a form of social participation, an educational experience and a factor for employability and social integration. It is essential for cities to create enabling environments for volunteering, training, empowering and rewarding volunteers and raising awareness on the role and value of citizens offering some of their priceless time to the city.

Cities which have organised and trained volunteers as a most precious urban human resource bear witness to multiple gains. Barcelona demonstrated the benefits of a city which has an organised voluntary body of passionate and committed citizens. In preparing for the Olympic Games, the city created and trained a body of 40,000 volunteers. After the Olympic Games, the local authorities decided to preserve this body as a living asset and offer it new opportunities. The municipality conferred prestige on these volunteers and helped the self-organisation of the body into the organisation "Volunteers 2000", which has been very active in many urban projects.

The Lille 2004 Cultural Capital of Europe was supported by over 17,800 citizens, residents or expatriates originating mostly from the Nord/Pas-de-Calais region and spread across the world. Present throughout the year, the ambassadors proved to be one of the driving forces of this exceptional event. In 2006, several hundred volunteers contributed to the success of Bombaysers de Lille.

In Athens, following the call for volunteers by the organisation Athens 2004, responsible for the 2004 Olympic Games, 160,000 international applications were submitted and 55,000 volunteers were selected and trained to offer their services during the Olympic Games. In a spirit of openness, the host city invited foreigners to participate on equal terms with local volunteers. The municipality of Athens also organised a dedicated body of volunteers to help visitors find their way throughout the city. This tradition continued for the preparation of the Athens 2011 Special Olympics. The president of Greece was the first citizen to apply to be a volunteer.

Respecting differences is essential for sustainable cities. Grass-roots organisations are most valuable for understanding urban diversity. They fight for better places, communities and planetary order. The "Our World Is Not for Sale" network, a loose grouping of organisations, activists and social movements in 70 countries, is committed on all fronts related to sustainable development.

Plurality, a key element in European civilisation, has played an important role in the development of cities and the formation of grass roots. Amsterdam, the European capital of tolerance, offers one of the best examples of openness. The first city in the world in the numbers of foreign-born citizens, it has always demonstrated its willingness to accept and integrate foreigners, exotic influences and new ideas into the city. Political refugees, intellectuals and workers of every race and religion have always lived side by side with the local population in a sober city, in which the power of no one person must be permanently felt anywhere (Pistor et al. 1994).

Many times the city balanced on the divide between harmony and anarchy, order and chaos. Yet, each time, the city managed to preserve its cultural diversity and dynamism and build its strength and resilience. The construction of the Amsterdam underground and the demolition of many old parts of the city, in the early 1970s, caused a peaceful protest by the local population of Nieuwmarkt, in the heart of the former Jodenbuurt. With a favourable public opinion and street art used as a weapon, residents obstructed the work of the demolition squads. The decoration of the completed metro station was assigned to the former demonstrators and the entire vault of the Nieuwmarkt station has been turned into a graphic account of the district's tumultuous history.

An unrivalled racial melting pot, the city also accepted that repression must not be tolerated. Institutionalised social movements were very active whenever the social fabric of the city was threatened in the 1960s and 1970s. In the 1980s, the aggravation of the housing problem caused violent clashes between the police and squatter communities evicted from buildings destined for high-profit use. Transport and housing improvements continue to be key issues and the object of referenda. Artificial islands have been created to provide both owner-occupied and rented housing for a wide range of income brackets, despite protests by ecologists about threats against biodiversity.

Citizens who are proud of the cities in which they live are a precious resource for any city. In Brussels, the Atelier de Recherche et d'Action Urbaines (ARAU) brings together aware citizens and local residents who care for their cultural, industrial and residential urban heritage. Since 1969, ARAU has called for high-quality developments in Brussels. The outrageous demolition of some important buildings, such as Horta's "People's Home" in the 1960s, spurred the unity of citizens willing to do something more than simply protest.

The challenge is to become responsible citizens and learn to read, live and write about the city, continuously shaped by historic, economic, social, cultural and political factors. Many associations organise guided tours, often combined with bicycle tours, offering alternative views of Brussels for citizens and visitors wishing to rediscover the often hidden richness of the city and reappraise urban life.

9.2   Institutional Innovation and Local Governance

Governance is the science and art of cogoverning societies with the participation of all actors. It is increasingly recognised that policy options cannot be based on artificial system management, but must be based on the evolving dynamics, will and preferences of society. In most policy areas, a new civic contract is being sought with civil society, which is expected to invigorate the debate between local, regional and national governments and the constituencies they represent and enhance public transparency, accountability and capacity for dialogue and decision. The political vigour of cities depends on enlightened leaders able to ponder fundamental forces and real opportunities and informed citizens able to judge among various possibilities according to their values and principles.

The cornerstones of the institutional architecture for sustainability are decentralisation and shared ownership of policies, together with an adequate transfer of powers and resources. Since the 1980s, governance systems in OECD countries have gone through considerable change, resulting largely from widespread decentralisation of government functions. This had a dramatic effect on policy making and led to the reallocation of tasks and resources and the development of more flexible institutional structures. A wide range of stakeholders, including private enterprises, the voluntary sector and the civil society, gradually constitute a new and more or less formal policy network, within which sustainability issues are discussed and policy options developed.

Dissensus being the source of multiple divisions, purposeful governance has to build a social consensus to sustain progress within stability. The most efficient and balanced allocation of functions among governmental and non-governmental bodies, both horizontally and vertically, as well as the best way to achieve it may be decisive in advancing towards sustainable development. To fulfil new mandates and ensure progress towards sustainable development, regional and local authorities should be endowed with the resources to manage the functions assigned to them. Fiscal federalism, based on the search for a balance between distribution of powers and allocation of resources, may be instrumental for sustainability.

Governance architectures for sustainable development increasingly include formal mechanisms of horizontal and vertical cooperation between governmental bodies and partnerships with non-governmental actors. Depending on the degree of decentralisation, local and regional authorities are creating the necessary institutional bridges among themselves, with the central government and with social partners and NGOs, to maximise local and regional participation in the process of policy formulation and implementation.

New spatiofunctional structures, such as intercommunal frameworks, regional platforms and territorial pacts, are promoted for both generic and specific purposes. They promise more coordinated spatial planning and more coherent allocation of public resources, as well as greater transparency, participation and accountability.

Stakeholders are sources of precious knowledge that can be harnessed to increase the responsiveness of public policy delivery. Over the past two decades, significant policy initiatives have been designed to promote empowerment, associative democracy and stakeholder involvement towards the shared ownership of projects and informed policy making.

In this context, the role played by negotiation and mediation in establishing new governance structures and in transforming inefficient relationships into dynamic partnerships is becoming increasingly prominent. Effective cooperation processes between various government departments, different tiers of government, the public and private sectors and voluntary and civil society actors can make a difference. Contract and negotiation processes have the potential to articulate the richness of information available at the local level with the broader national vision. They may lead to better and more responsive policies and to a more effective and accountable allocation of resources.

Leadership at all levels is necessary to provide a coherent vision and policy framework, enable territorial authorities to set priorities and invest resources, promote public–private partnerships and share the risk and cost of innovations. Participative leadership at the local and regional levels is necessary to define territorial needs and develop strategies and programmes, mobilise public and private resources and develop a permanent dialogue with the other territorial authorities and stakeholders.

Stockholders, the owners of the physical assets, stakeholders, having particular interests in local projects, and other partners, shareholders providing resources and competencies, may invest great energy in cities and open new perspectives for the future. Strategic public–private partnerships have a great potential in reinforcing the objectives of short-term economic vigour with long-term societal objectives. Policies that promote collaboration with the business sector facilitate firm networking and clustering and transfer of cleaner technologies are extremely important. The success of these approaches depends on the overall policy environment, encompassing both macroeconomic and structural conditions.

Metropolitan areas face special challenges for institutional reforms and new models of governance, enhancing their human, social, natural, environmental and human-made capital. The OECD principles of metropolitan governance highlight the importance of promoting coherence, competitiveness, coordination, equity, fiscal probity, flexibility, holistic and adapted approaches, participation, subsidiarity and sustainability (OECD 2000b).

To meet the challenges of the new millennium, Toronto devised a civic laboratory of institutional change. After having benefited from a particularly successful two-tier governance model for over 40 years, the regional government and its six municipal authorities amalgamated into a single unified city of Toronto. Institutional structures tried to reposition the city for sustainable development and foster a democratic and diverse urban community, socially cohesive and environmentally healthy (OECD-Toronto 1997).

The role of governance in integrated planning can be demonstrated by the Umlandverband Frankfurt (Greater Frankfurt Development Agency), a public corporation serving the central area of the Rhine/Main region. The agency is responsible for the overall development of the two major conurbations of Frankfurt and Offenbach and 41 other municipalities. It also acts as a regulatory body on issues related to municipal and environmental planning, water supply and waste management (EEA 2009c).

The preventive Environmental Atlas is a major tool created by the agency. The countless data on the nature of environmental media are presented as maps of environmental strains, refined on various scales to support enhanced decision making by administrations and citizens. Although the permissible strains are defined at national level, the consultation of local actors is essential as they bear much responsibility concerning the pressures on the environment. The agency also created maps with the environmental impact on the ecosystems, highlighting the problems stemming from the synergies of different strains placed on the environment.

In 2009, Brussels celebrated 20 years of the Brussels-Capital region. As a result of constitutional and administrative reform in Belgium in the late 1980s and the 1990s, Brussels was established as one of three autonomous regions of Belgium, along with the Flemish and Walloon regions. With little more than one million inhabitants in 2008 and 300,000 foreigners, the Brussels-Capital region organised a consultation process for the future of an urban agglomeration wishing to be a true capital of Europe and accelerate change towards sustainable development (Bruxelles Capitale-Agence de Développement Térritorial 2009).

9.3 Rights and Duties to the City: Citizenship, Participation and Accountability

The evolution from a representational system of democracy to a more interactive participatory democracy is occurring in many regions and countries. Grass-roots movements want direct contacts between the governing and the governed. Representative democracy faces the challenge of the duly constituted authorities, linked to the representative role of local groups. Democracy should provide forums to exert sound judgement and help citizens to be transformed from mere consumers into city actors, sharing values, visions and actions (Abbott 1996).

Governments at all levels should democratise expertise and reinforce legitimacy, accountability and transparency. An EC study highlights that access to sound information and knowledge, transparency and accountability, effectiveness, early warning and foresight, plurality, independence and integrity have to reinforce scientific excellence. Technology foresight must be accompanied by social forward-looking activities, to support both better-quality decision making and public trust in the scientific foundation of policies. Urban projects and programmes must not only be scientifically robust, but also socially acceptable.

The search for sustainable development introduced a multiplicity of complex issues, spanning a variety of actors and domains. Demonstration projects and communication should enhance the participation of the underrepresented social groups and open the decision-making process. Only an organised society can form a strong basis for resource preservation and sustainable development. Making the community, especially the underrepresented and vulnerable social groups, more knowledgeable, aware and willing to participate in the management of the common goods is an important task for local governments. It is not unrelated to scientific progress and technological diffusion.

Sustainability offers cities the opportunity to become new democratic spaces between the world macroregulations and the microregulations of the local communities. Cities must create a dynamic environment in which economic prosperity, social cohesion and citizenship can blossom. Cities should be able to inject vision into this environment and enable all actors to take control of their common future.

Citizenship involves everyday participation in all aspects of the city. Citizens are increasingly invited to act as partners rather than protesters. Many urban projects, ranging from the improvement of exceptional vernacular architecture to the tracing of new metro lines, have been crowned with success thanks to the extraordinary active participation of responsive citizens (EFILWC 1996a, 1998c).

Cities and regions with a strong tradition of civic engagement are often more open to innovation and change. The example of Emilia-Romagna, a region with traditional openness and a constellation of small and medium-sized enterprises, is significant. Putnam (2002) charted the powerful influence of citizen participation in this region to buttress institutional performance. He asserted that the region is not populated by angels, but that the social capital and the cultivation of the civic community promote strong and responsive collective action.

Local democracy is reinforced from an everyday reconfirmation of civic values, an ongoing reinforcement of the civic bond. Bologna was the first European city to organise a referendum on the closing down of its historic centre to private cars. In Brussels, the consultation procedures for planning introduced new participation concepts. Reggio Emilia invited citizens to participate in the compiling of the city budget. In Amsterdam, various referenda sought the population's views on issues ranging from the new metro line to the extension of the residential areas of the city.

The channels of citizen representation and participation have been enriched with new instruments and methods. The Barcelona "Civic Agreement 2002" is a partnership between the city and the main citizen associations. The charette method is being used to bring together the richness of diverse opinions and ideas and build communication while projects are still on the drawing board. Action planning and workshop schemes involve the organisation of carefully structured collaborative events, which liberate creative individuals, articulate a sense of purpose and create a momentum, a thrust for the future. From London to Moscow, action planning weekends have helped develop visions and strategies for urban renaissance and sustainability.

Action planning schemes, dialogue and consensus workshops can bring together many different, traditionally opposed, actors, on neutral ground and on equal terms on the broad spectrum of sustainable development issues. Mirror groups, sounding boards, citizens' juries and other intermediary platforms can provide more permanent and effective interfaces among experts, policy-makers and citizens (Healey 1997).

Forward-looking national, regional and local governments create strategic public–public partnerships, enriched with effective public–private partnerships. Public–private partnerships have the potential to reduce the social costs of projects and lead to better synergies among public and private investments. They offer ample grounds for coalitions to overcome thematic and institutional barriers and play a critical role in the implementation of sustainable development policies.

Public–private partnerships are linked to the shift in public policies from direct interference to indirect or conditional policies, such as incubation and mediation. Partnerships should work like an orchestra (private) with its conductor (public) and produce music and not noise. Innovative partnerships can optimise the potential of synergies, enrich content and methods of cooperation and act as catalysts of change.

Partnerships usually build upon the shared values and priorities of a city and the commitment of societal actors to act upon common interests. Experience suggests that to be successful, partnerships need a clear vision and structure, a strategic and tactical approach, a critical mass, assertive leadership and social justice, continued evaluation and assessment. Partnerships should enhance the capacity, contribution and commitment of the public, private and community sectors and fortify the social capital of a city.

The passionate involvement of local communities to improve their world can make a great difference. The Finglas Enlivenment Project, on the periphery of Dublin, in the early 1990s, is an inspiring one. The area included extensive public housing estates, built cheaply to accommodate inner-city slum clearances and population growth. The Finglas planning team, set up to synchronise forward-planning and development functions, identified local key individuals and groups and promoted consultation to enhance the positive aspects of the area. The first project initiated with the local chamber of commerce targeted the improvement of public spaces, through a shopfront competition and the redesign of the historic centre. Impressive results were achieved with modest funds. A sculpture offered by a local artist, composed of models of the hands of all the citizens of Finglas, serves as an allegory for the project (EFILWC 1996a).

Lille, a pioneer city in promoting dialogue and consultation, can count on citizen participation in regular debates held in the City Hall. As early as 1978, Lille was one of the first cities in France to create district forums, some 25 years before they became a legal obligation for all towns with a population of over 60,000. The Euralille urban development project, centred on the new TGV station, has fostered a long debate among citizens. The metropolis enhanced its experience as 2004 Cultural Capital of Europe, as a step further in involving citizens in designing its future.

9.4 A World Pact for Cities and Citizens: For a Global Coalition of Excellence

City-twinning schemes have planted the seeds for the creation of a world compact for cities. The earliest examples of twinning cities include the treaties between ancient city states designed to protect each other's interests in times of hostilities. The Romans were particularly effective at such politics and the tradition continued into the Renaissance. Most recent twinning schemes in Europe have their origins in the hope of peace and the unprecedented involvement of the citizenry in warfare. As early as 1944, Vancouver established such a relationship of solidarity with Odessa.

The twinning activity has been much more intense since the end of the Second World War. In 1947, Bristol, for instance, sent leading citizens to Hanover and Edinburgh and signed a twinning agreement with Nice. And on the other side of the Atlantic, President Eisenhower instituted the so-called Sister City programme in 1956 which became a citizen diplomacy network creating and strengthening partnerships to increase global cooperation at the municipal level, promote cultural understanding and stimulate economic development. The programme created a movement for local community development and volunteer action by motivating and empowering private citizens, municipal officials and business leaders to engage in long-term programmes of mutual benefit.

The choice of twin cities may be based on various geographical, industrial or cultural factors, growing from long-standing traditions rooted in the past or recent cultural or other links prompted by political sentiments of solidarity, such as with Poland in the 1970s, and more recently with China and Cuba. Old university towns have close links, such as Leiden twinned with Oxford and Cambridge twinned with Heidelberg.

The latest EU enlargements brought together cities and towns which once belonged to countries lying across the old dividing line of the Iron Curtain. An imaginative example comes from Vienna and Bratislava, which created the "Twin City Vienna-Bratislava". Their partnership may lead to one of the most influential urban agglomerations in central Europe. Vienna has a thriving high-technology biotechnology industry, whereas Bratislava has attracted many car manufacturers. The new Twin City has the best of opportunities to become, as it was 100 years ago, the heart of central European economic life.

The Partners of the Americas, a voluntary organisation linking the citizens of 46 US states with the people of Latin America and Caribbean countries, is another example of twinning at the level of the semicontinent. Founded in 1964, the Partners of the Americas links volunteers in many areas of expertise to build cultural awareness, respect, improve living and working conditions and build leaders in their respective communities. San Francisco has long served as a partner city to Mexico City and supported economic and cultural exchanges beneficial for citizens on both sides of the border. Bay Area residents worked with a counterpart committee in Mexico City and volunteered to teach English, bring medical supplies and develop microenterprises.

At the microlevel, twinning of urban neighbourhoods can go even further in linking people. Downtown Bonn is surrounded by a number of districts with a history of their own and which were independent municipalities until a few decades ago. As many of these had their contacts and concluded their own twinnings before the local government reorganisation, Bonn entertains a tight-knit network of twinnings with municipalities in other European countries. Bonn Innenstadt ("Inner City") maintains twinnings with Oxford and a district of Budapest. Bad Godesberg, the ancient resort town, today a district of Bonn, is twinned with St. Cloud in France, Frascati in Italy, Windsor-Maidenhead in England and Kortrijk/Courtrai in Belgium. The District of Beuel on the right bank of the Rhine and the District of Hardtberg cultivate twinnings with the French towns of Mirecourt and Villemomble, respectively. All of these microtwinnings feature a lively interchange in the spirit of international understanding.

On the global scene, city alliances could be instrumental for achieving the Millennium Declaration specific goals for 2015. The review of progress against achievements highlighted the main positive results but also underlined that much remains to be done (OECD 2005).

The last of the UN millennium goals concerns the development of a global partnership for development. The social compact among developing countries, doing more to ensure their development, and developed countries, supporting them through aid, debt relief and trade access, has been beneficial but further commitments are needed. Progress is also required on the Doha round, technology transfer, access to essential care, and promotion of youth employment. Cities could play a great role.

The Clinton Climate Initiative (CCI) joined efforts with cities in a business-oriented approach to fight against climate change in practical, measurable and significant ways. The C40 Large Cities Climate Leadership Group comprises cities in the developed world, developing world and emerging economies and their alliance seems decided and able to make a difference.

Municipal governments are well positioned to make a major impact on the global market for green technologies. Cities purchase building materials, appliances and systems for thousands of buildings, such as schools, hospitals, offices and police stations. Cities also buy and operate municipal fleets of vehicles and run their water and waste systems. Through their partnership with CCI, cities can collectively pioneer energy-efficient and clean-energy products and technologies.

Joint public procurement can significantly reduce greenhouse gas emissions and mitigate climate change on a large and measurable scale.

In May 2007, the creation of the global Energy Efficiency Building Retrofit Programme brought together eight of the world's largest energy service companies, five of the world's largest banks and 17 of the world's largest cities in a landmark programme designed to reduce energy consumption in existing buildings. The programme provides both cities and their private building owners with access to the necessary funds to retrofit existing buildings with more energy efficient products, typically leading to energy savings between 20 and 50%. If all participating cities were replacing all appliances in municipal buildings with energy-efficient ones, much energy could already be saved.

As part of the Clinton Global Initiative, Amsterdam is cooperating with Seoul, San Francisco and Cisco through the Connected Urban Development programme for making cities more sustainable. Part of the programme is the development of urban eco-maps to create awareness among citizens of the impact of carbon emissions on their urban environment. It provides information on carbon emissions from transportation, energy and waste among neighbourhoods, organised by district, and gives tips on ways to reduce a resident's carbon footprint. Such tools providing emissions information on a neighbourhood level, organised by local and comparable common boundaries such as ZIP codes and districts, could empower citizens to move from collective intelligence to collective action.

The development of urban infrastructures and technology transfer to the developing world could be two domains of intercity cooperation with a higher potential for sustainable development. They require effective cooperation between governments, industry and financial institutions. They could also offer a business perspective and interest world coalitions of innovative businesses. Developing world cities could be the gates for the cleaner technologies to upgrade the world infrastructure for water, energy and sanitation. Significant penetration of cleaner technologies in developing countries would need access to the best available technology, capital for the necessary investments and an adequate institutional and financial framework. Cities, probably in cooperation with business coalitions, could facilitate smooth international exchanges, access to effective markets and services and transfer of state-of-the-art technology.

The picture of the developing world cities is very complex and diverse. On energy infrastructure, the example of cities in some OPEC countries (Venezuela, Algeria, Nigeria) which built their entire economy on oil transit and exportation is significant. The structure of energy supply in the developing world as a whole presents some particular features. There is relatively more use of coal and renewables, particularly owing to the primitive use of biomass. Non-commercial traditional energy, mainly including firewood, charcoal, crop residues and animal waste, accounts for approximately 10% of global primary energy use, but reaches 30% in developing countries and 70% of final energy consumption in sub-Saharan Africa. Its use, based on obsolete technologies, is highly unsustainable and, in many cases, also contributes to deforestation. Fuel-switching represents an opportunity for cleaner energy options and international cooperation should help this be realised.

Cities in the EU, the world's largest donor of development aid, have a critical role to play, particularly in creating strategic partnerships for the transfer of cleaner technologies. Advanced energy technology dissemination actions, carried out in cooperation with cities in developing countries, could boost the development of energy services based on renewable sources.

The penetration of renewable technologies into the developing megacities of the world can yield multiple benefits for the environment, security of energy supply and the global economy. Decentralised generation presents a special opportunity and the command and control power grids, which presently leave a large share of the population outside, could gradually be completed with more decentralised and efficient, cleaner and service-oriented systems.

Cities could get inspiration from IEA collaborative projects, known as "Implementing Agreements", which offer the mechanisms and structures for the development of new energy systems, as well as for the deployment of clean technologies in the market place. For more than 30 years, activities under the Implementing Agreements have been fundamental building blocks in facilitating progress of new or improved energy technologies. The legal contracts and a system of standard rules and regulations allow interested member and non-member governments or other organisations to pool resources and to foster the research, development and deployment of particular technologies. The Technology Agreement on End-Use Technologies/Buildings provides a good example.

Buildings and infrastructures are the most tangible assets that cities have in common and in a much more concentrated form than the rest of the world. Much energy is still used inefficiently in buildings which could be transformed into net energy producers. Energy efficiency is a top priority on the international agenda. The deployment of energy-efficient equipment is the most cost-effective immediate path to sustainable development, greater energy security and lower greenhouse gas emissions. The IEA estimates that energy-efficiency improvements could contribute 47% of reductions in energy-related CO_2 emissions potentially achievable by 2030. Cities could facilitate the drastic decrease of the inefficiency of buildings. Cooperation could extend to domains of industrial and traffic technologies. Cities could create a global compact and move the world forward.

Watercolour 10
Athens: The Birthlight of Urban Democracy

Athens: The Birth light of urban democracy

Chapter 10
Sustainability Indicators:
The Benchmarks of Urban Excellence

Sustainability indicators can take the pulse of a city and provide insight regarding progress towards sustainable development. They may serve as a yardstick and be instrumental for policy making and for assessing policy implementation... But, as Albert Einstein formulated, "not everything that can be counted counts and not everything that counts can be counted." Qualitative indicators are often necessary to complement quantitative frameworks.

This chapter insists on the importance of a policy framework for the development of indicators and presents some influential practices with indicators and targets. It finally proposes a set of headline policy-significant environmental, economic and sociocultural indicators for cities, in line with the principles of the European Charter of Sustainable Cities.

10.1 Robust and Significant Indicators To Assess and Compare Progress

Sustainability indicators and indices may serve as compasses on the journey to urban sustainability. They are useful tools for evaluating, reporting and reorienting progress towards sustainable development. Retrospective indicators and appraisal targets can help assess policy implementation, whereas prospective indicators can be instrumental for policy making. The question of the measures is critical. Qualitative indicators can be as important as the quantitative indicators they complement.

Indicators are conventionally defined as performance measures that aggregate information into a useful and usable form. They should be meaningful, clear, scientifically sound, i.e. verifiable and reproducible, and accepted, objects of debate and consensus. Descriptive indicators, illustrating the status of a city and based on real, concrete physical measures, are easier to establish and interpret by judging them against specified benchmarks. Performance indicators, based on policy principles and goals, have to be integrated in a policy framework, including a diagnosis of the situation, identification of the factors that should change and directions and

V.P. Mega, *Sustainable Cities for the Third Millennium: The Odyssey of Urban Excellence*, DOI 10.1007/978-1-4419-6037-5_10,
© Springer Science+Business Media, LLC 2010

actions for the desired change. Ultimate targets to be attained and the process for monitoring implementation should ideally complement the policy framework.

Indicators are not simple statistical data but intellectual constructs that try to capture, in the best possible way, certain complex and evolving realities. Their significance extends beyond what is directly obtained from observations. The "pressure–state–response" model is a widely accepted framework for the construction of sustainability performance indicators. The model links the causes of changes (pressure) to their effects (state) and finally to the projects, actions and policies (response) which are designed and implemented.

Typical wealth indicators do not capture aspects related to the environment and quality of life. GDP, the most used measure of macroeconomic activity, developed in the 1930s, cannot support policy debates on sustainable development. The move beyond GDP is an implicit recognition that the market cannot fulfil all aspirations, especially when, under the effects of crisis, there is a strong need for public intervention, from the local to the global level. An overall consensus is building up to move towards more adequate indices of progress, wealth and well-being, including environmental and social indicators (EC 2009g, Stiglitz 2009). Already in the 1970s, Bhutan, the first country to profit, in 2009, from the Adaptation Fund, created under the UNFCCC, helping poor countries to adapt to climate change, proposed the "Gross Domestic Happiness indicator".

Sustainability indicators have often been linked to the question of greening the national accounts systems, aggregating and comparing data in monetary forms. Genuine savings indicators or gross welfare product indices attempt to broaden the usual measure of saving (equal to the difference between income and expenses) to account for the cost of environmental depletion and degradation and investment in human capital or welfare (OECD 2000a).

The Human Development Index (HDI) combines normalised measures of life expectancy, educational attainment, and GDP per capita for countries worldwide. It is claimed as a standard means of measuring human development, a concept that, according to the UN Development Programme, refers to the opportunities of people for education, health care, income, employment, etc. The HDI attempts to measure a country's development by combining life expectancy at birth, adult literacy rate and standard of living, as measured by the natural logarithm of GDP per capita at purchasing power parity. According to the HDI 2009, the countries with the highest HDI include Norway, Australia, Iceland and Ireland (UNDP 2009).

The UN Commission on Sustainable Development proposed in 1996 a method and a framework of indicators, organised according to the "driving force–state–response" model (UN 1996). Indicators for supporting and promoting sustainable human settlement development form part of this framework. They comprise three driving force indicators (rate of growth of urban population, per capita consumption of fossil fuel by motor vehicle transport, human and economic loss due to natural disasters), state indicators (percentage of population in urban settlements, area and population of urban formal and informal settlements, floor area per person and house price to income ratio) and one response indicator (infrastructure expenditure per capita).

European efforts to create a common set of urban indicators have multiplied over the last decade. The EU sustainability monitoring initiative "Towards a local sustainability profile - EU common indicators" launched in 2000, has been developed as a bottom-up approach in close collaboration with local authorities. Communities participate on a voluntary basis. The set includes five core indicators on citizen satisfaction with the local community, contribution to global warming, mobility patterns, green spaces and local services and air quality. Five additional indicators have been suggested. They focus on children's commuting patterns, local management, noise pollution, sustainable land use and sustainable products, including eco-labelled, organic and fair-trade products.

Indicators should capture critical features of a city and contribute to making the city more visible and transparent, help structure and harmonise databanks, provide decision-makers with relevant and timely information, assist appraisals, comparison and prediction, stimulate communication and promote citizen empowerment and participation. They should embrace all sectors and neighbourhoods contributing to the coevolutionary process of sustainable development. An indicator assessment board should validate the set of indicators, after they have been tested, but the framework should remain living and be regularly updated.

Urban thematic indicators should reveal a city's performance in all fields contributing to sustainable development and according to the specific policy objectives. An aggregate index, such as the HDI, the genuine savings indicator or the ecological footprint, may inform about the overall performance of a city. The passage from thematic indicators to an index of sustainability policy performance for cities is a complex task, since indicators have to be weighted by the contribution to sustainability levels and all the previous levels of aggregation have to be taken into account. Special effort must be made to avoid multiple counting of individual sustainability pressures, which are taken into account in the composition of the thematic indicators. Finally, it is important to highlight that no indicator can inform if a city integrates socioeconomic and environmental policy objectives in its overall development strategy (OECD 1996).

Targets for thematic indicators may be defined at the city level, according to the priorities of each city, with probable reference to global protocols, supranational and national commitments and standards. The performance of a city at a national or supranational level should therefore be judged according to its local objectives, its national objectives and its international commitments.

Many cities have tested and introduced frameworks of indicators during the last few years. Seattle, in the USA, is often quoted as a classic example of a dynamic city, a breeding ground of successful businesses such as Boeing and Microsoft, with a coherent set of indicators. Last but not least, indicators should reinforce communication processes and promote common understanding at the local, regional, national and global levels (Sustainable Seattle 2008).

The Boston Indicators Project issued its first report in 2000 with the goal of tracking incremental progress until 2030, Boston's 400th anniversary. "The Wisdom of Our Choices: Measures of Progress, Change and Sustainability" report introduced a framework of indicators and measures identified through a rigorous process involving more than 300 experts and stakeholders. The report noted that the booming knowledge economy was creating a divide between those with and without a good education. The second report "Creativity and Innovation: A Bridge to the Future" (2001–2002) highlighted Boston's institutional, physical and cultural assets, but noted as a trend to watch the move of young people away from Boston and Massachusetts due to high housing costs and other factors.

The next biennial report "Thinking Globally, Acting Locally: A Regional Wake-Up Call" (2003–2004) noted that the region was suddenly competing for jobs and talent not only with other US regions, but also with China, India and other emerging economies. It called for a coherent, collaborative response and issued an Emerging Civic Agenda. The fourth biennial report on the period 2005–2006, during which the local and regional economy strengthened measurably, suggested remarkable progress had been made on the civic agenda.

The Boston Indicators Project offers new ways to understand Boston and its neighbourhoods in a regional context. It aims to democratise access to information, foster informed public discourse, track progress on shared civic goals, and report on progress in ten sectors: civic vitality, cultural life and the arts, economy, education, environment, health, housing, public safety, technology and transportation. The Web site of the Boston Indicators Project allows access to data for more than 200 measures of 70 shared civic goals. The 2004–2006 Boston Indicators Project report illustrates trends until 2006 (Boston Foundation 2005, 2003, 2001).

International comparative analyses should always be regarded with a sound dose of scepticism, since they are limited by national differences in definitions and data collection and estimation methods. Comparisons are meaningful when they refer to truly comparative, in size and function, units. Systematic territorial indicators are necessary complements to national indicators serving for international comparisons. Moreover, reporting regularly and systematically on territorial progress towards international targets and commitments, focuses attention on the task ahead and national decision-makers become more accountable at the local and the international level.

The SustainLane US City Rankings of the 50 largest cities in the USA constitutes an inspiring benchmark exercise on the unfolding efforts of cities towards sustainable development. Since the first SustainLane US City Rankings in 2005, world events have made sustainability an even more vital concept for US cities. Hurricanes Katrina and Ike illustrated the vulnerability of urban dwellers and the 2008 crisis underlined dependency on unpredictable market forces. SustainLane's rankings cover indicators of quality of life, such as local food availability, air and water quality, pedestrian and park space and roadway congestion. SustainLane's city rankings also track the growth of clean technologies, developments in renewable energy, waste management, advanced transport services, alternative fuels and green buildings.

The SustainLane US City Rankings focus on urban policies and practices which differ across the nation and the way this affects citizens. The winner of the third edition was Portland, ranked as the most sustainable city in 2009, with a good to great performance in most fields. Ranked below average only in affordability, natural disaster risk and water supply, Portland excels in clean technology and green building development, overall quality of life, and sustainability planning and management. More than ever, citizens suggest that they have a high quality of life and work hard to be involved in city policy, boards, projects and practices that impact sustainability.

In Portland, leadership for sustainability is exercised at all levels. It is not only demonstrated by the local governments but also by all actors, including the grass roots and the businesses. The municipal administration provides an example by promoting renewable power and green buildings. Citizens opt for the most sustainability oriented options in everyday life. The builders, developers and buyers are the ones who change the market. Portland continues to use its sustainability ethos to attract businesses, residents and tourists to become one of the capitals of a powerful emerging domestic economy.

The EU Urban Audit, launched by the European Commission in 1997, constitutes a major exercise to develop indicators. Its purpose is to enable an assessment and benchmarking of EU cities. Several concepts were tested and large volumes of data were collected during the pilot study in 1999, the large-scale data collection round of 2003–2004 and the collection round of 2006–2007. The data, which went through stringent quality controls, have been available from Eurostat since April 2008. The added value of the Urban Audit lies in its wide choice of indicators, its large geographical coverage and its long time series.

Fifty-eight cities were invited by the European Commission to participate in the Urban Audit during the pilot phase. The data collection expanded to cover 258 cities in 2003–2004 and 367 cities in 2006–2007, 321 from the 27 member States, 26 Turkish cities, six Norwegian cities and four Swiss cities. The cities were selected in cooperation with the national statistics offices and are geographically dispersed to ensure a representative sample. Data are collected for the core cities, for the larger urban zones and for a small subset of variables focusing on districts.

The audit consists of more than 300 indicators, composed and calculated from the 336 variables collected by Eurostat. They cover most aspects of quality of life, e.g. demography, housing, health, crime, labour market, income disparity, local administration, educational qualifications, environment, climate, travel patterns, information society and cultural infrastructure. Data availability is domain-specific and webmetrics increasingly enrich data series.

Following the first Urban Audit of 2003–2004, a report on the state of European cities offered a wide range of conclusions. The strongest urban population growth rates were recorded in peripheral EU15 member States and particularly in Spain, where some urban areas saw average annual increases of 2% or more. Cities in Ireland, Finland and Greece also experienced some of the highest population growth rates in the EU. In contrast, many urban areas in central and eastern Europe witnessed an overall population decline in the same time frame. In virtually all cities, suburbs grew and if they declined, they still tended to decline less than the core city (EC 2007f).

In 2009, for the first time, 185 cities from all 27 EU member States were covered by the Urban Atlas, produced by the European Commission and member States with the support of European space technology. Composed of thousands of satellite photographs, the Urban Atlas provides detailed and cost-effective digital mapping, ensuring that city planners have the most up-to-date and accurate data available on land use and land cover. The Urban Atlas will enable urban planners to better assess risks and opportunities, ranging from the threat of flooding and the impact of climate change, to identifying new infrastructure and public transport needs. The Urban Atlas covers all EU capitals and a sample of large and medium-sized cities participating in the European Urban Audit. All cities in the EU will be covered by the Urban Atlas by 2011.

Apart from the real facts and figures, perception studies are important complements to indicator frameworks. The Urban Audit Perception Survey, conducted in November 2006 to assess the local perceptions of quality of life in 75 cities in the EU27, Croatia and Turkey, complements the data from the main Urban Audit exercise. The perception survey was conducted as a Flash Eurobarometer survey. All capitals were included and in the larger member States one to six more cities were added. In each city and in all parts of the cities, 500 randomly selected individuals were contacted and asked 23 questions about the quality of life in their city. Overall, citizens declared being satisfied with their quality of life. However, the answers to specific questions express much more diverse feelings, including some strong dissatisfaction. For example, in many cities, it is not easy to find a good job or it is difficult to find good housing at a reasonable price.

The Urban Atlas, complementing the Urban Audit, demonstrates the benefit of an integrated European approach and is an excellent example of how space-based technology applications contribute to local solutions across Europe. The maps provide a pan-European classification of city zones, allowing for comparable information on the density of residential areas, commercial and industrial zones, the extent of green areas and watercourses, exposure to flood risks and monitoring of urban sprawl.

10.2   Principles for a Meaningful Set of Urban Indicators

Agenda 21 called for countries, international organisations and non-governmental organisations to develop and use indicators of sustainable development. Cities have to develop indicators linked to their Local Agenda 21 plans. Sustainability indicators cannot include solely environmental indicators, as environmental performance is not the only factor in achieving sustainability. The framework has to be completed with appropriate socioeconomic indicators.

Local Agenda 21 and the Charter of European Cities and Towns may serve as valuable policy frameworks for the development of performance indicators. An indicator can be a priori assigned for each policy theme. The differences in the nature and the scale of the policy themes and the diversity of territories dictate the variety of the indicators to be developed. The indicators that are composed should assert whether a city follows the adopted directions for sustainability and should shed light on the actual results or the remaining steps needed to be taken to incorporate irreversibly into sustainable pathways. Frameworks and sets of indicators underline the importance of focusing on the policy issue and the objectives.

The indicative set of headline urban indicators which is proposed in the next section was conceived following the Charter of European Sustainable Cities and Towns (ESCTC 1994). The framework includes policy-significant environmental, economic and sociocultural indicators, while assigning each city a unique sustainability indicator.

The environmental indicators include both indicators representing local environmental quality and the contribution to national and global environmental change. The indicators expressing the territorial responsibility for the global environment (responsibility for global climate, acidification of the environment, toxification of ecosystems) and local disturbances follow the directions of the Dutch set (Adriaanse 1993). The air quality indicator, urbanisation, sustainable mobility, energy and water consumption indicators and resource management indicators relate to crucial aspects of the local environment linked to global considerations.

The economic dimensions of sustainability are encapsulated by the composite indicator of economic sustainability. It is composed from the territorial income, fiscal deficit, environmental expenditure and pollution damage. The indicator of social sustainability was composed from access to employment, income, information, education and training for all citizens. The sustainable buildings indicator has been defined in a way to capture the quality of both housing and working conditions and sustainable construction aspects for existing and new buildings, whereas public safety and citizen participation indicators were chosen to express other crucial aspects of sociopolitical sustainability. Next to these, the quality of green, heritage and public space indicator serves as a measure of the quality of spaces promoting public health, social life and cultural identity. Finally, for each city, it would be useful to assign a unique sustainability indicator representing the contribution of specific local assets, characteristics or events to sustainability.

The scale for the development of each policy-significant indicator is very much dependent on its nature: global climate, acidification, ecosystem toxification or economic sustainability indicators may be more significant at a regional level, whereas other indicators ,e.g. on resource management or nuisances, may be more relevant at a more local level.

Subindicators are very important in revealing specific features or helping achieve specific targets. For example, the per capita consumption of water or resources is very important when individual responsibility is involved and personal behaviour patterns have to be addressed. Subindicators are also important for identifying gender gaps.

The set of urban indicators which is proposed was initially developed by the author in the framework of a thesis on sustainability indicators at Harvard University and served as the basis for the development of indicators by various cities, notably in the framework of a project managed by the European Foundation in Dublin (EFILWC 1998a). During the last 10 years, the set of indicators has been improved by the author (Mega 2005, 2008) and benefited from her involvement in international panels discussing and assessing indicators, the last of which was convened by The Economist Intelligence Unit in Frankfurt in April 2009.

The proposed framework of indicators can be enhanced by indicators such as GDP per capita, energy consumption of transport, resource productivity, healthy life expectancy, risk of poverty, rate of employment of older workers, official assistance to development and the effectiveness of political consistency. Especially in times of crisis it is not advisable to include intensity indicators (e.g. energy intensity indicator, equivalent to energy per unit of GDP) because reduced GDP growth may give ambiguous signals on increased nominators.

10.3   Indicative Set of Headline Urban Indicators

1. Climate change indicator

Definition	The contribution of a city to global climate change.
Policy objectives	Decrease of greenhouse gas emissions, according to national and international commitments.
Components	Emitted quantities of CO_2, CH_4, N_2O, CFCs and halons.
Measure	Climate change equivalent: total greenhouse gases (CO_2, CH_4, N_2O and CFCs).
Composition	The indicator is the sum of the quantities of the greenhouse gases, weighted according to their warming potential. The degree (N) to which each greenhouse gas contributes to global warming depends on its concentration in the troposphere and on its ability to absorb the heat radiated by Earth.
Policy portfolios	Urban policies for buildings, urbanisation, green areas, energy, transport and resource management can have an important impact. Targets for the various gases can be more ambitious than the national targets; monitoring is critical.
Remarks	*The target can be composed from the targets for the emissions of each of the constituent greenhouse gases.*

2. Air quality indicator

Definition	The number of days per year on which local alarm levels defined by law are exceeded in the most negative measurement of SO_2, CO, NO_x, ground-level ozone and particulate matter emissions.
Alternative definition	*SO_2, CO, NO_x, ground-level ozone and particulate matter emissions.*

Policy objectives   Improvement of air quality for all, through the reduction
                        of SO_2, CO, NO_x, ground-level ozone and particulate
                        matter emissions, according to local objectives and
                        commitments.

Subindicator   Number of days per year on which alarm levels are
                        exceeded and special traffic measures have to be
                        introduced.

Policy portfolios   A city has to identify and manage all sources of SO_2,
                        CO, NO_x, ground-level ozone and particulate
                        matter emissions so that the levels of air quality are
                        acceptable by the standards of the WHO.

3. Acidification indicator

Definition   The deposition of acidic components.

Measure   Acidification equivalents: the total acidification caused by
                        acidic compounds and deposited per hectare.

Policy direction   Drastic reduction of deposition of acidic compounds.

Components   Deposition of SO_2 per hectare, deposition of NO_2 per
                        hectare and deposition of NH_3 per hectare.

Policy portfolios   A city has to identify all sources of acidic compounds
                        which may have an impact and ensure that depositions
                        are minimal.

4. Ecosystem toxification indicator

Definition   The emission of hazardous toxic substances.

Measure   Toxic substances equivalent: the total emission of priority
                        substances and radioactive substances.

Policy direction   Reduction of the quantity of each of the hazardous
                        substances released by a given territory to a level
                        where the risk posed by each substance is negligible.

Components   Emitted quantities of cadmium, polyaromatic
                        hydrocarbons, mercury, dioxin, epoxyethane, fluorides
                        and copper. Emitted radioactive substances.

Composition   The indicator is the sum of the emitted quantities
                        of priority and radioactive substances, weighted
                        according to their toxicity and their residence time
                        in the environment.

Policy portfolios   A city has to identify all sources of toxic substances and
                        ensure that their emissions are negligible.

5. Sound urbanisation indicator

Definition   The progressive urbanisation of a city.

Measure	Sustainable urbanisation equivalent: the percentage of green areas in the greater agglomeration that are converted each year into urban residential areas.
Policy direction	Consolidating the urban fabric and reducing urban sprawl.
Subindicators	Construction indicators for greenfields compared with generation indicators for brownfields.
Components	Construction indicators for greenfields/brownfields.
Policy portfolios	Measures for reducing urban sprawl and consolidating cities include the creation of mixed urban quarters well served by public transport.

6. Sustainable mobility indicator

Definition	The use of environmentally friendly means of transport, especially for commuting and fundamental mobility needs. The term "sustainable mobility" may be considered as an oxymoron, a contradiction in terms. "Less unsustainable mobility" could be a more precise notion.
Measure	Sustainable mobility equivalent: the total number of passenger kilometres by environmentally friendly means (foot, bicycle and public transport) per inhabitant and per year. If the number of passenger kilometres cannot be estimated, the number of trips could be an approximation.
Policy direction	Reduction of the unnecessary use of motor vehicles, improvement of mobility patterns and the modal split and enhancement of accessibility.
Subindicators	Sustainable mobility for commuting (work and study) indicator. Relevant subindicators may be developed according to other trip purposes (business, freight, tourism, leisure) and in relation to transport means. A "sustainable mobility for children" indicator could be most interesting.
Components	Total number of trips (and their length) by each transport means with special focus on the most sustainable transport means (by foot, bicycle and public transport).
Remarks	*The length of pedestrian areas and cycle paths over time is also a useful indication of actions to promote less unsustainable mobility.*
Policy portfolios	Measures to enhance sustainable mobility include the provision of pedestrian and bicycle paths, banning of private cars in city centres, congestion charges and municipal fleets running on fuel cells and hydrogen.

7. Resource (and waste) management indicator

Definition	The degree to which natural resources (and waste) are used and finally transformed into solid waste.
Measure	Resource (and waste) management equivalent expressed in tonnes of finally disposed waste per inhabitant and per year.
Policy direction	The primary aim is solid waste minimisation and optimisation, through prevention and avoidance, followed by reuse, remanufacturing and recycling; drastic reduction of waste for final disposal.
Subindicators	Waste enhanced and reused, recycled and finally disposed of (ideally zero).
Components	Building and demolition waste, industrial, domestic waste, and commercial and service waste (overall and per capita).
Composition	The indicator is the sum of all waste streams. The ultimate target should be zero waste.
Remarks	*The indicator considers only solid waste. Liquid waste can be distinguished by the degree of treatment. Possible overlap with ecosystem toxification should be considered.*
Policy portfolios	Promote full cycle resource management and advance towards zero waste.

8. Energy consumption indicator

Definition	The total amount of energy to satisfy the demand of a city and its impact on the environment.
Measure	Energy consumption equivalent expressed in tonnes of oil equivalent per year.
Policy direction	Reduction of energy consumption through improved energy-efficiency techniques and patterns. Decrease of energy dependency and energy originating from polluting sources and progress towards decentralised renewable energy production.
Subindicators	Consumed energy according to the origin (coal, oil, natural gas, nuclear energy, wind, hydro, solar and other renewable sources), weighted according to its impact on the environment.
Components	Energy to satisfy the final demand by the residential sector (and per capita consumption), the tertiary sector, the industrial sector and transport.
Composition	The energy consumption indicator is the total amount of energy needed to satisfy the demand and its impact on the environment.
Policy portfolios	Improve energy efficiency, promote renewable energy sources and cleaner technologies, and capture and sequestration of greenhouse gas emissions.

9. Water consumption indicator

Definition   The total amount of water withdrawal.

Measure   Water consumption equivalent expressed in cubic metres
                      per year.

Policy direction   Reduction of water consumption through minimisation of
                      water losses in pipelines, improved conservation patterns
                      and techniques; reuse and recycling.

Components   Water for domestic purposes (and per capita), industrial use,
                      the building sector, retail services and public spaces.

Composition   The water consumption indicator is the total amount of
                      water extracted. Water from recycling, used mainly
                      for maintenance of public and green spaces is to be
                      subtracted.

Remarks   *The quantity of water lost in the mains should be estimated
                      (5–40% of the total amount). Water management represents
                      the highest environmental expenditure in OECD countries and
                      minimisation of water losses is essential.*

Policy portfolios   Improve water quality, radically diminish water losses in
                      pipelines and generate water savings, including through
                      pricing of water and the use of recycled water for cleaning
                      purposes.

10. Local nuisance indicator

Definition   Local nuisances due to noise, odour or visual pollution.

Measure   Local nuisance equivalent: percentage of the population
                      affected by noise (e.g. from air, road and rail traffic and
                      industry), odour (e.g. caused by traffic, industry and
                      services), or visual pollution (e.g. by derelict land and
                      social degradation).

Policy direction   Improvement of local environments by reduction of odour,
                      noise or visual pollution.

Subindicator   It is essential to have a subindicator for the percentage of
                      the population adversely affected by one of the above-
                      mentioned factors.

Components   Percentage of the population affected by noise, odour, or
                      visual pollution.

Composition   The total number of people affected is the sum of the people
                      affected by any one of these sources, after correction has
                      been made to avoid nuisances from overlapping sources, as
                      simultaneous exposure to different types of nuisance may occur.

Policy portfolios   Measures to monitor and reduce noise, odours that may cause
                      disagreement and visual pollution.

11. Social sustainability indicator

Definition	The degree of social sustainability of a city, expressed as access to employment, income, information, education and training for all citizens.
Measure	Social sustainability equivalent: the percentage of people not affected by poverty, unemployment, homelessness and lack of access to education and training.
Policy direction	Access to employment, income, housing, education, information and training.
Subindicators	Gender subindicators are most important for social sustainability. It is essential to have a subindicator for the percentage of the population seriously affected by the lack of each of the above-mentioned components. It is also important to have subindicators for key population groups (youths, handicapped, long-term unemployed, immigrants, elderly, older old etc.).
Components	Percentage of the population affected by poverty, unemployment, homelessness and lack of access to education, information and training.
Composition	The indicator can be composed after subtraction of the total percentage of people affected by social exclusion, the sum of the percentages of people affected by poverty, unemployment, homelessness and lack of access to education, information and training.A corrective factor has to adjust the percentage of the population affected by more than one factor.
Policy portfolios	Active prevention of social exclusion and unemployment, measures for employment creation, facilitation of access to housing, social assistance to affected citizens, especially the ones near the poverty line, and provision of information, education and training.

12. Sustainable buildings indicator

Definition	The sustainability performance of buildings for housing and for public, commercial, service and industrial use buildings.
Measure	Sustainable buildings equivalent: the percentage of people enjoying adequate housing and working environments with the least impact on the environment.
Policy direction	Offer all inhabitants the possibility to live and work in healthy buildings with the least impact on the environment.
Subindicator	Sustainability performance of residential buildings, further distinguishing them into passive and active buildings, buildings for public use and commercial, tertiary and industrial buildings.Sustainability performance of existing/new buildings.

Components   Population living in active ecological buildings; population
                  working in active ecological buildings; percentage of
                  ecologically sound public, commercial, tertiary and
                  industrial buildings; percentage of ecologically sound
                  public and industrial buildings.

Composition   Population enjoying adequate housing and working conditions
                  with the least impact on the environment or percentage of
                  sustainable buildings compared in the overall building stock.

Remarks   *The number of registered demands for active ecological
                  buildings is an indication of the actual demand and may
                  serve as an alternative indicator.*

Policy portfolios Measures for fewer unsustainable energy-efficient buildings
                   and a clean housing and working environment for all,
                   including provision of information, training and advice to
                   reduce the environmental footprints of buildings.

13. Public safety indicator

Definition   The degree to which people enjoy public safety.

Measure   Public safety equivalent: the total percentage of the population
                  enjoying public safety.

Policy direction Fostering public safety for all citizens. Decrease in, ideally
                  elimination of, attacks, incidents and accidents.

Subindicator   It is essential to have subindicators for the total percentage of
                  irreversible long-term injuries and for vulnerable groups
                  of population (youth, women, elderly, the handicapped and
                  long-term unemployed).

Components   Percentage of people attacked and percentage of people
                  affected by road accidents.

Composition   The indicator can be composed after subtraction of the total
                  percentage of citizens affected by lack of safety.

Policy   Measures to improve safety of public places, including
  portfolios   telesupervision and technological improvement of vehicles
                  and modern safety systems for protecting private areas
                  without barriers.

14. Economic sustainability indicator

Definition   The viability of the territorial economy.

Measure   Economic sustainability equivalent: territorial income minus
                  territorial fiscal deficit minus environmental expenditure
                  minus pollution damage per inhabitant per year.

Policy direction Increase of economic sustainability with increase of territorial
                  income and budget and reduction of pollution damage.

Components   Territorial income (total individual incomes), territorial fiscal
                      deficit (territorial budget minus taxes), environmental
                      expenditure (for water, waste collection, sewage, transport)
                      and pollution damage (air, water and land).

Composition   As for "Measure" above.

Remarks   *The territorial environmental expenditure per inhabitant per*
                      year is a good indicator of the local financial capacity and
                      the concern about the environment.

Policy portfolios   Measures to improve economic sustainability, including local
                      energy, water and waste taxes.

15. Green, public and cultural space indicator

Definition   The percentage of high-quality green, public and cultural/
                      heritage spaces accessible to all citizens.

Measure   Green, public and cultural space equivalent:the percentage
                      of high-quality green or/and public or/and cultural spaces
                      accessible to all citizens.

Policy direction   Enhancement of green, public and cultural spaces, forging
                      cultural identity and belonging. Ensuring accessibility of
                      high-quality spaces to all.

Subindicators   Alternative indicators include the area of accessible high-
                      quality green space per inhabitant, the area of cultural/
                      heritage space per inhabitant and the area of public space
                      per inhabitant.

Components   Percentage of green spaces in need of improvement per total
                      area of green space, the percentage of public spaces
                      in need of improvement per total area of public
                      space and the percentage of cultural/heritage spaces
                      in need of improvement per total area of cultural/heritage
                      space.

Composition   The total percentage of accessible high-quality green, public
                      and cultural spaces.

Policy portfolios   Measures to improve green, public and cultural/heritage spaces.

16. Citizen participation indicator

Definition   The engagement of the local population in the decision-
                      making for urban sustainability.

Measure   Citizen participation equivalent: the total percentage of the
                      population participating in local elections or as active
                      members in local associations.

Policy direction   Cogoverning cities in partnership with all societal actors.
                      Behavioural change is also a critical dimension of citizen
                      participation.

Components   Percentage of people participating in local elections;
                  Percentage of people being active members of
                  environmental, public health and cultural associations.

Composition   The total percentage of the population active in local elections
                  and participating in associative life.

Remarks   *It is important when assessing the participation in local*
                  elections to know if participation in the elections is
                  obligatory or not. In the EU, voting is compulsory in
                  Belgium, Greece and Luxembourg.

Policy portfolios Measures for local engagement of citizens include information
                  and communication campaigns and awareness raising.

17. Unique sustainability indicator

Definition   An indicator to be defined case by case according to the
                  uniqueness of a city (e.g. unique climatic and local
                  conditions) or the planning of a unique once-in-a-lifetime
                  high-impact event (e.g. Olympic Games, Expos). This
                  indicator should represent the degree to which unique
                  factors or events may lead to sustainability.

Policy portfolios Measures taken by the city for it to benefit in the long run
                  from an ephemeral event.

Watercolour 11
Cities of the Mind/Homaga to P. Klee

Voula P. Mega graduated as an engineer from the National Technical University of Athens and completed her *diplôme d'études approfondies* (DEA) at the National Geographical Institute in Paris. She continued with a DEA at the French Institute of City Planning, where she was also conferred with a Ph.D. degree in city and regional planning.

Her postdoctorate studies include research on regional policy at Oxford Brooks University and environmental economics and policy analysis at the Harvard Institute for International Development.

She started her career as Special Adviser to the Greek Government. She has been Research Manager at the European Foundation, an EU agency in Dublin, and worked as an expert on sustainable development at the Organisation for Economic Cooperation and Development (OECD) in Paris. She joined the European Commission in Brussels in 2001 and presently works as a policy officer at the Research Directorate-General.

She has written, co-written and edited more than 30 books and 100 articles published by the EU, UN, UNESCO, OECD and international publishers and magazines, in Greek, English, French, Spanish and Italian. The most recent titles include *Sustainable Development, Energy and the City* (Springer Science+Business Media, New York, 2005) and *Modèles pour les Villes d'Avenir* (L'Harmattan, Paris, 2008). Her main subjects include innovation and sustainable development, urban dynamics, energy and transport, regional capital, enterprise and the environment, city and spatial policy and cultural added value.

She also published the poetry collections *Siren Cities* (in Greek, English and French; Exantas, Athens, 1997) accompanied with drawings (Exantas, Athens, 1997) and *Dawns and Souls for Europe/Aubes et Âmes pour l'Europe* (in English and French; Persée, Paris 2008), an official selection for the European Book Prize 2008. An exhibition of watercolours, drawings, diptychs (with poems) and photographs at the European Parliament (February 2004) united her artistic work on cities under the title "The Song of Siren Cities".

Selected Web Sites

The purpose of this list is to assist readers who navigate into various websites; they may need to add the correct Internet protocol (e.g. http://) and the correct capitalization (web site addresses are case-sensitive, depending on which type of Web server, e.g. Linux/UNIX or Microsoft, on which they are hosted)

Cities in an Era of Fragility

- WWW.EC.EUROPA.EU/ENVIRONMENT/EUSSD/
- WWW.EUROPA.EU/COMM/REGIONAL_POLICY/URBAN
- WWW.EUROPA.EU/COMM/ENVIRONMENT/URBAN
- WWW.SUSTAINABLE-CITIES.ORG
- WWW.AALBORGPLUS10.DK
- WWW.EOLSS.NET
- WWW.UN.ORG/ESA/SUSTDEV
- WWW.UN.ORG/ERA/UPP
- WWW.UN.ORG.PP
- WWW.UNFPA.ORG
- WWW.UNAIDS.ORG
- WWW.URBACT.EU
- WWW.WORLDWATCHINSTITUTE.ORG
- WWW.WHITEHOUSE.GOV/ISSUES/URBAN_POLICY/

For a New Balance Among Nature, Humans and Artefacts in Cities

- HTTP://CALCULATOR.BIOREGIONAL.COM/
- WWW. EUROPA.EU/COMM/ENVIRONMENT/
- WWW.EUROPA.EU/CLIMATE ACTION
- WWW.ICLEI.ORG
- WWW.IPCC.CH
- WWW.C40CITIES.ORG

- WWW.UNFCCC.INT
- WWW.EEA.EU.INT
- WWW.FOOTPRINTNETWORK.ORG
- WWW.PANDA.ORG/LIVINGPLANET
- WWW.LEGRENELLE-ENVIRONNEMENT.FR
- WWW.UNEP.ORG/URBAN_ENVIRONMENT/
- WWW.UNHABITAT.ORG
- WWW.UGEC.ORG
- WWW.CLIMATESUMMITFORMAYORS.DK
- WWW.LONDON.GOV.UK
- WWW.DOGME2000.DK, WWW.KK.DK
- WWW.ACRPLUS.ORG
- WWW.MYCLIMATE.ORG
- WWW.TORONTO.CA/BBP/
- WWW.WASTEENG.ORG
- WWW.HAMBURG.DE, STOCKHOLM.SE

Transport for Sustainable Cities

- WWW.EC.EUROPA.EU/TRANSPORT
- WWW.ECMT.ORG WWW.INTERNATIONALTRANSPORTFORUM
- WWW.CIVITAS-INITIATIVE.EU
- WWW.NICHES-TRANSPORT. ORG
- WWW.SENIORCITE.RATP.FR
- WWW.TRANSPORTBENCHMARKS.ORG
- WWW.FUBICY.ORG
- WWW.TTR_LTD.COM
- WWW.UITP.COM
- WWW.VELO-CITY2009.COM
- WWW.VELIB.FR, VILLO.BE

Energy for Sustainable Cities

- WWW.EUROPA.EU/COMM/ENERGY
- WWW.ADEME.FR
- WWW.EREC.ORG
- WWW.ENERGIE-CITES.ORG
- WWW.ERE-RENEWABLES.ORG
- WWW.EWEA.ORG
- WWW.ISES.ORG, ESTIF.ORG
- WWW.FUELCELL-INFO.COM
- WWW.MANAGENERGY.NET
- WWW.RENEWABLES2004.DE/EN/2004

- WWW.SUSTENERGY.ORG
- WWW.SAVE MORETHANFUEL.EU
- WWW.TOTNES.TRANSITIONNETWORK.ORG
- WWW.WINDDAY.EU
- WWW.ASIMPLESWITCH.COM

Competitive and Attractive Cities

- WWW.WEFORUM.ORG
- WWW.KOBENHAVN.DK
- WWW.STOCKHOLM.SE
- WWW.IBA.DE
- WWW.AIVP.ORG
- WWW.CICI.ORG
- WWW.MERCERHR.COM
- WWW.CUSHWAKE .COM
- WWW.BEIJING2008.CN

Social Justice and Solidarity in Cities

- WWW.FAS.IE
- WWW.SOCIALEXCLUSIONUNIT.GOV.UK
- WWW.WHO.DK/HEALTHY -CITIES
- WWW.URBANSECURITY.ORG
- WWW.FESU.ORG
- WWW.LASCE.FR
- WWW.UNION-HABITAT.ORG
- WWW.SOCIALEDGE.ORG
- WWW.VILLE.GOUV.FR
- WWW.ESPOIR-BANLIEUES.FR/
- WWW.YOUNGFOUNDATION.COM

The Cities of Sciences, Culture and the Arts

- WWW.INTELLIGENTCOMMUNITY.ORG/
- WWW.CITIESOFSCIENCE.CO.UK
- WWW.SCIENCE YEAR.COM
- WWW.CITE-DES-SCIENCES.FR
- WWW.SCIENCE-ET-CITE.CH
- WWW.EC.EUROPA.EU/CULTURE
- WWW.EUROPANOSTRA.ORG

- WWW.ICOMOS.ORG
- WWW.UNESCO.ORG/CULTURE
- WWW.RUELIBRE.FR
- WWW.PRINTEMPSDESEPTEMBRE.COM

The Urban Renaissance of the Third Millennium

- WWW.GLASGOW.GOV.UK
- WWW. CITIES21.COM
- WWW.BCN.ES
- WWW.BERLIN.DE
- WWW.HARBOURBUSINESSFORUM.COM
- WWW.TEMPOMAT.IT
- WWW.URBANVILLAGES.COM
- WWW.URBAN-VILLAGES-FORUM.ORG.UK
- WWW.URBANTASKFORCE.ORG
- WWW.BIENNIALTOWNPLANNING.ORG

The Cities of the Citizens

- WWW.EC.EUROPA.EU/CITIZENSHIP
- WWW.ARAU.ORG
- WWW.COE.INT/CLRAE
- WWW.C40CITIES.ORG
- WWW.4D.ORG
- WWW.COMITE21.ORG
- WWW.CLINTONFOUNDATION.COM
- WWW.WISEREARTH.ORG
- WWW.UNHABITAT.ORG

Sustainability Indicators: The Benchmarks of Urban Excellence

- WWW.EC.EUROPA.EU/REGIONAL_POLICY/UR BAN2/URBAN/AUDIT
- WWW.BEYOND-GDP.EU
- WWW.SUSTAINLANE.COM/US-CITY-RANKINGS/
- WWW.BOSTONINDICATORS.ORG
- WWW.SUSTAINABLESEATTLE.ORG
- WWW.VEOLIA.FR
- WWW.MONITORINGDESQUARTIERS.IRISNET.BE

Watercolour 12:
Bangkok: The Temple of the Dawn

Annex: Twenty International Networks and Movements of Cities

At the world level:

Large Cities Climate Leadership Group, also known as the *C40 Cities* (and originally as the C20 Cities) is a group of the world's largest cities (Addis Ababa, Athens, Bangkok, Beijing, Berlin, Bogotá, Buenos Aires, Cairo, Caracas, Chicago, Delhi, Dhaka, Hanoi, Hong Kong, Houston, Istanbul, Jakarta, Johannesburg, Karachi, Lagos, Lima, London, Los Angeles, Madrid, Melbourne, Mexico City, Moscow, Mumbai, New York, Paris, Philadelphia, Rio de Janeiro, Rome, São Paulo, Seoul, Shanghai, Sydney, Tokyo, Toronto and Warsaw) committed to tackling climate change. Through effective partnership with the Clinton Climate Initiative, C40 strives for better energy efficiency and emissions reductions. The current chair of the C40 is the Mayor of Toronto.

The *International Council for Local Environmental Initiatives* (ICLEI) is an international association of local governments and regional organisations that have made a commitment to sustainable development. It brings together over 1,105 cities, towns counties, and their associations worldwide to work together on international performance-based, results-oriented campaigns and programmes.

The *Sustainable Cities Programme* is a joint programme of UNEP and the UN Human Settlements Programme (UN-HABITAT), which supports local authorities in urban environmental planning and management. It was established in the early 1990s to build capacities and was founded on broad-based stakeholder participatory approaches. Currently the Sustainable Cities Programme and its sister programme Localising Agenda 21 (LA21) operate in over 30 countries worldwide.

Cities Alliance (Cities Without Slums) is a coalition of partners through which UNEP supports cities in poverty reduction and sustainable development.

The international organisation of *United Cities and Local Governments* (UCLG) was launched in Paris in 2004 as the result of a merger between the International Union of Local Authorities (IULA), the World Federation of Unitied Cities (FMCU-UTO) and Metropolis. Its mission is to be the united voice and world advocate of local governments, through the promotion of their values, objectives and interests.

The *Association of Cities and Regions for Recycling and Sustainable Resource Management* (ACR+) is an international network of some 100 members in

20 countries with the common objective of sustainable waste and resource management. ACR+ members are essentially local and regional authorities, along with national networks of local authorities representing some 1,100 municipalities.

The *Global Cities Dialogue* (GCD) is a worldwide network of cities, created in 1999, promoting an information society free of digital divide and based on sustainable development. Key interlocutors are the political authorities representing the cities and acting as driving forces for change processes and dialogue at bilateral or multilateral level.

The *Organisation of World Heritage Cities* brings together 198 cities that are home to an inhabited site featured on UNESCO's World Heritage list. It is a nongovernmental international organisation which aims to promote the implementation of the World Heritage Convention and the International Charter for the Conservation of Historic Towns.

Metropolis is the world association of major metropolises. Metropolis is also the metropolitan section of the UCLG, which arose from the merger between the IULA and UTO. Created in 1985, Metropolis is represented by more than 100 member cities from across the world and operates as an international forum for exploring issues and concerns common to metropolises.

The *International Association of Educating Cities*, founded in 1994, brings together local governments wishing to cooperate on projects and activities to improve the quality of life of the citizens. It has more than 300 city members in 31 countries.

United Cities Against Poverty, initiated in 2001 by the cities of Bamako (Mali), Geneva (Switzerland) and Lyon (France), is an association of cities and local authorities whose aim is to promote cooperation by developing and implementing sustainable and fair local public policies, with an emphasis on solidarity.

The *Worldwide Network of Port Cities* (AIVP) is an international network of public and private stakeholders implicated in the sustainable development of harbour cities. It brings together elected representatives of cities and other local governments, maritime, and waterway bodies, port administrations, operators, enterprises, service providers, universities and research institutes.

The *Union Internationale des Transports Publics (UITP)* is a worldwide network of public transport professionals from all means of transport (underground, bus, light rail transit, etc.). It has more than 2,700 local, regional and national members from more than 80 countries from the five continents.

The *World Mayors' Council on Climate Change*, initiated by the then mayor of Kyoto upon entry into force of the Kyoto protocol, is an alliance of committed local government leaders advocating an enhanced recognition and involvement of local governments in multilateral efforts addressing climate change and related issues of global sustainability. The council is chaired by the former mayor of Bonn.

At the European level:

The Council of European Municipalities was founded in Geneva in 1951 by a group of European mayors; later, it opened its ranks to the regions and became the *Council of European Municipalities and Regions* (CEMR). It is the largest organisation of local and regional governments in Europe, with more than 50 national associations of towns, municipalities and regions from 37 countries. Together these associations represent some 100,000 local and regional authorities.

EUROCITIES is the network of major European cities. Founded in 1986, the network brings together the local governments of more than 130 large cities in over 30 European countries. It provides a platform for its member cities to share knowledge and ideas, to exchange experiences, to analyse common problems and to develop innovative solutions, through a wide range of forums, working groups, projects, activities and events.

The *WHO European Healthy Cities Network* consists of a network of cities from around Europe that are committed to health and sustainable development. It comprises over 1,200 cities and towns from over 30 countries in the WHO European Region, linked through national, regional, metropolitan and thematic healthy cities networks.

Énergie-Cités is an association of European local authorities promoting local sustainable energy policies. It was created in 1990 and represents more than 1,000 towns and cities in 26 countries. From 2009 to 2011, Énergie-Cités is under the presidency of Heidelberg.

The *Climate Alliance of European Cities with the Indigenous Rainforest Peoples/ Alianza del Clima e.V.* is Europe's largest city network for climate protection. It works for the preservation of the global climate, through reducing greenhouse gas emissions in the industrialised countries of the North and preserving the rainforests in the South of the planet.

MEDCITIES is a network of Mediterranean coastal cities created in Barcelona in 1991 on the initiative of the Mediterranean Technical Assistance Programme (METAP). The METAP, whose objective is environmental improvement in the Mediterranean region, was established in 1990 by the World Bank, the European Investment Bank, the European Commission and the UNDP.

The *Union of the Baltic Cities* (UBC) is a network mobilising over 100 member cities for democratic, economic, social, cultural and environmentally sustainable development of the Baltic Sea Region. Its aims and goals are stated in the UBC Statute and UBC Strategy.

References

Abbott, J. 1996. Sharing the City: Community Participation in Urban Management. London: Earthscan

ACDHRD (Australian Commonwealth Department of Housing and Regional Development). 1995. Urban Futures. Canberra

ACR+. 2009. Municipal Waste in Europe - Towards a European Recycling Society. Paris: Victoires

Adriaanse, A. 1993. Environmental Policy Performance Indicators. Dutch Ministry of Housing, Physical Planning and Environment, The Hague

Ajuntament de Barcelona. 1995. Barcelona: un modelo de transformación urbana 1980–1995. Barcelona: Gestion Urbana, Vol. 4

ALFOZ. 1995. La Ciudad accesible. No. 109. Madrid

Amsterdam Climate Office. 2008. New Amsterdam Climate – Summary of Plans and Ongoing Projects. Amsterdam

Architects' Council of Europe (ACE) (The). 2009. Survey on the Impact of the Economic Crisis on Europe's Architectural Profession. Brussels

Arene (Agence régionale de l'environnement et de nouvelles énergies). 2007. Les ateliers de pratique urbaine en Europe. Quartiers durables en Europe

Aristotle. 350 BC. Politics. Translated by Benjamin Jowett (1950). Oxford University Press

Aristotle. 350 BC. Nichomachean Ethics. Translated by Terence Inwin (1985). Ed. Hackett

Beatley, T. 2000. Green Urbanism: Learning from European Cities. Washington, DC: Islands Press

Bellet, C et al. 1998. Ciudades Intermedias. Urbanizacion y sostenibilidad. Lleida: Milenio

Berlin capitale. 1992. Berlin Capitale. Un Choc d'identités et de Cultures. Berlin

Boston Foundation. 2007. A Time Like no Other. Chartering the Course of the New Revolution. A Summary of Boston Indicators 2004–2006

Boston Foundation. 2005. Thinking Globally, Acting Locally: A Regional Wake-Up Call. Boston

Boston Foundation. 2003. Creativity and Innovation: A Bridge to the Future. Boston

Boston Foundation. 2001. The Wisdom of Our Choices: Measures of Progress, Change and Sustainability. Boston

Bremen Municipality. 1997. The Bremen Declaration. Business and municipality: New Partnerships for the 21st Century. Bremen

Brown A. et al. 2008. Urban Policies and the Right to the City: Rights, responsibilities and citizenship. UNESCO (MOST Programme). Paris

Brown, L.R. 2006. Plan B 2.0. Rescuing a Planet Under Stress and a Civilisation in Trouble. New York: Norton & Co

Bruxelles Capitale-Agence de Développement Territorial. 2009. Bruxelles 20 ans. Brussels

BURA. 1997. The Future of Cities. Conference Report. Belfast

Burwitz et al. 1991. Vier Wochen ohne Auto. Bericht über ein freiwilliges städtisches Abenteuer. University of Bremen

Castells, M. 2001. The Internet Galaxy. New York: Oxford University Press

Castells, M. and Hall, P. 1994. Technopoles of the World. The Making of twenty- first century industrial complexes. London: Routledge

CERES. 2008. Ceres Sustainability Report. Boston

CERES. 2006. Corporate Governance and Climate Change: Making the Connection. Boston

Chesneaux, J. 1996. Habiter le temps. Passé, présent, futur : esquisse d'un dialogue politique. Paris : Bayard

Clinton, W. 2007. Giving. How Each of us can Change the World. New York. A. Knopf

Comité 21. 2008. Agir ensemble pour des territoires durables ou Comment réussir son Agenda 21. Paris

Council of Europe. 2008. La Charte urbaine européenne II – Manifeste pour une nouvelle urbanité. Strasbourg

Council of Europe. 1992. The European Urban Charter. Strasbourg

Council of European Municipalities and Regions (CEMR). 2009. The Economic and Financial Crisis: Impact on Local and Regional Authorities. Paris

Council of Ministers. 2008. Déclaration finale des ministres en charge du développement urbain. Marseilles

Council of Ministers. 2007. Leipzig Charter on Sustainable European Cities. Leipzig

Cushman and Wakefield. 2009. European Cities Monitor. London

Cushman and Wakefield. 2007. Main Streets Across the World 2007. London

Cushman and Wakefield. 2006. Office Space Across the World. London

Danzig, G.B. & Saaty T.L. 1973. Compact City: A Plan for a Liveable Urban Environment. San Francisco: Freeman

Delanoë, B. 2008. De l'audace. Paris: Lafont

De Portzamparc, C. 2007. Rêver la ville. Paris: Le Moniteur

De Portzamparc, C. 1996a. Disegno e forma dell'architettura per la città. Rome: Oficina

De Portzamparc, C. 1996b. Généalogie des formes. Paris: Dis Voir

De Portzamparc, C., Sollers, P. 2003. Voir Ecrire. Paris : Calmann -Lévy

DIV. 2009. Réussir un projet de dévelopement urbain. Paris

DIV. 2004. Zones franches urbaines: Une chance à saisir ensemble. Paris

DIV. 2003. Talents de cités. Paris

DIV. 1995. Territoires urbains et exclusion sociale. Paris

Doxiadis, C. 1975a. Building Eutopia. Athens Publishing Centre

Doxiadis, C. 1975b. Action for Human Settlements. Athens Publishing Centre

Doxiadis, C. 1974. Anthropopolis. Athens Publishing Centre

EKISTICS. 2002. Defining success of the City in the 21st century. Vol 69. No 415/416/417. Athens

Ecological Footprint network et al. 2008. Living Planet Report. Oakland

Economist (The). 2007. The World Goes to Town. 4 May 2007

Economist (The). 1995. Survey on Cities: Turn on the Lights. 29 July 1995

ENA Recherche. 1996. La Ville et ses usagers. Paris : La documentation française

Énergie – Cités. 2001. Database for Municipal Good Practice. Brussels

Enerpresse. 2006. Bâtiment : Le défi énergétique. Paris : Le Moniteur

ERRAC. 2004. Light Rail and Metro Systems in Europe. Brussels

Eurocities. 2009. Strategies Against Homelessness. Brussels

European Commission. 2009a. Sustainable Development: a Challenge for European Research. Conference Proceedings. Brussels

European Commission. 2009b.Climate Change: Act and Adapt. Green Week Brussels, 23-26 June 2009. Conference Proceedings. Brussels

European Commission. 2009c. The 2009 Review of the EU Sustainable Development Strategy. Brussels

European Commission. 2009d. A Sustainable Future for Transport: Towards an Integrated Technology-led and User-Friendly Transport. Brussels

European Commission. 2009e. The World in 2025. Rising Asia and Socio-Political Transitions. Brussels

European Commission. 2009f. The Recovery Plan. Brussels

European Commission. 2009g. GDP and Beyond Measuring Progress in a Changing World. Brussels

European Commission. 2008a. La réalité sociale en Europe. Brussels

European Commission. 2008b. Europe's energy position. Present and Future. Market Observatory for Energy 2008. Europe. Brussels

European Commission. 2008c. Sustainable Energy Europe. Showcasing Europe's Best Energy Solutions. Brussels

European Commission. 2008d. Attitudes Towards Radioactive Waste. Brussels

European Commission. 2008e.EU Environment – Related Indicators 2008. Brussels

European Commission. 2007a. An Energy Policy for Europe. Brussels

European Commission. 2007b. Limiting Climate Change to 2° Celsius: The Way Ahead for 2020 and Beyond. Brussels

European Commission. 2007c. Green Paper for Urban Transport. Brussels

European Commission. 2007d. Progress Report of the European Sustainable Development Strategy. Brussels

European Commission. 2007e. *Villes à vivre*. In Research EU, magazine de l'espace européen de la recherche, Nr 54. Brussels

European Commission. 2007f. The State of the European Cities Report. Brussels

European Commission. 2006a. Thematic Strategy on Urban Environment. Brussels

European Commission. 2006b. InfoRegio No 19. Cities for growth, jobs and cohesion. Brussels

European Commission. 2006c. Special Eurobarometer 258. Energy Issues. Brussels

European Commission. 2005a. European Values in the Globalising World. Communication. Brussels

European Commission. 2005b. Cohesion Policy and Cities: the Urban Contribution to Growth and Jobs in the Regions. Brussels

European Commission. 2004a. Environment for Europeans. No 17. Brussels

European Commission. 2004b. Sustainable Consumption and Production in the European Union. Brussels

European Commission. 2003a. Partnership with the Cities. The URBAN Community Initiative. Brussels

European Commission. 2003b. Second European Climate Change Programme. Brussels

European Commission. 2002. Science and Technology for Sustainable Energy: EU Research Visions and Actions.

European Commission. 2001. EU Strategy on Sustainable Development. Brussels

European Commission. 2000. Green Paper: Towards a European Strategy for the Security of Energy Supply. Brussels

European Commission. 1999. Metropolis 2000. A Conference Report. Brussels

European Council of Town Planners (ECTP). 1998. The New Charter of Athens. Athens

European Environment Agency (EEA). 2009a. Annual EC Greenhouse gas Inventory 1990–2007 and Inventory Report 2009. Copenhagen

European Environment Agency. 2009b. Environmental Signals. Copenhagen

European Environment Agency. 2009c. Ensuring Quality of Life in Europe's Cities and Towns. Copenhagen

European Environment Agency. 2007a. Europe's Environment. The Fourth Assessment. Copenhagen

European Environment Agency. 2007b. Exceedance of Air Quality Limit Values in Urban Areas. Copenhagen

European Environment Agency. 2006. Urban Sprawl in Europe. The Ignored Challenge. Copenhagen

European Environment Agency. 2005. The European Environment. State and Outlook 2005. Copenhagen

European Environment Agency. 2002. Towards an Urban Atlas: Assessment of Spatial Data on 25 European Cities and Urban Areas. Copenhagen

European Forum for Urban Safety (EFUS). 2007. Secucities Against Terrorism. Paris

European Forum for Urban Safety. 2006. The Cities' Manifesto for Safety and Democracy. Paris

European Forum for Urban Safety. 2003. Secucities, Schools and Cities. Paris

European Foundation for the Improvement of Living and Working Conditions (EFILWC). 1998a. Urban Sustainability Indicators. Dublin

EFILWC. 1998b. Challenges for Urban Infrastructures in the EU. Dublin

EFILWC. 1998c. Challenges for Urban Governance in the EU. Dublin

EFILWC. 1997a. Perceive – Conceive – Achieve the Sustainable City. Dublin

EFILWC. 1997b. European Cities in Search of Sustainability. A Panorama of Urban Innovations in the European Union. Dublin

EFILWC. 1997c. Utopias and Realities of Urban Sustainable Development. Dublin

EFILWC. 1997d. Medium-Sized Cities in Europe. Dublin

EFILWC. 1997e. Towards an Economic Evaluation of Urban Innovations. Dublin

EFILWC. 1997f. Redefining Concepts, Challenges and Practices of Urban Sustainability. Dublin

EFILWC. 1997g. Innovative and Sustainable Cities. Dublin

EFILWC. 1996a. Intermediate Cities in Search of Sustainability. Dublin

EFILWC. 1996b. What Future for the Urban Environment in Europe. Contribution to HABITAT II. Dublin

European Photovoltaics Industry Association. 2001. Solar Electricity in 2010. Brussels

European Renewable Energy Centres (EUREC) Agency. 2002. The Future for Renewable Energy. Prospects and Directions. London: James & James

European Sustainable Cities and Towns Campaign (ESCTC). 1994. Charter of European Cities and Towns: Towards Sustainability. Brussels

European Wind Energy Association. 2001. European Wind Energy Conference and Exhibition. Working Documents. Copenhagen

Eurostat. 2007. Measuring Progress Towards a more Sustainable Europe. 2007 monitoring Report of the EU Sustainable Development Strategy. Brussels

Florida, R. 2008. Who 's your city. New York: Basic Books

Florida, R. 2005. Cities and the Creative class. Routledge

Florida, R. 2004. The Rise of the Creative Class. New York: Basic Books

Foster. 1997. Invention, Innovation and Transformation. Cambridge: MIT Centre for International Studies

Fussler, C. 1996. Driving Eco-innovation. London: Pitman

Friends of the Earth. 1995. Towards a Sustainable Europe. Amsterdam

Galbraith, J.K. 1996. The Good Society: The Human Agenda. Boston. Houghton Mufflin Company

Garreau, J. 1991. Edge City. Life on the New Frontier. New York: Doubleday

Garvey, J. 2008. Ethics for Climate Change. London: Continuum

Gelford, P., Jaedicke, W., Winkler, B. and Wollmann, H. 1992. Ökologie in den Städten. Basle: Birkhäuser

Gillo, B. and Solera, G. (ed.) 1997. Sviluppo Sostenibile e Città. Naples: Clean

Gore, A. 2007. An Inconvenient Truth. New York: Adapted for a New Generation.

Greenpeace. 2005. Decentralising Power: An Energy Revolution for the 21st Century. London

Hahn, E. 1997. Local Agenda 21 and Ecological Urban Restructuring. A European Model Project in Leipzig. Berlin: WZB

Hall, P. 2002. Cities of Tomorrow: An Intellectual History of Urban Planning and Design in the Twentieth Century. London. Blackwell

Hall, P. and Pfeiffer, U. (ed.) 2000. Urban Future 21: A Global Agenda for 21st Century Cities, London: Taylor & Francis

Hall, P. 1998. Cities in Civilisation. Culture, Technology, and Urban Order. London: Weidenfeld & Nicolson

Hall, P. 1995. The European City: Past and Future. In The European City. Sustaining Urban Quality. Copenhagen

Hall, P. 1983. The World Cities. London. Weidenfeld & Nicolson

Harvard University. 1994. HIID Executive Course on Environmental Economics. Cambridge

Harvey, D. 1983. Social Justice and the City. London

Hewitt, M. 2001. City Fights: Debates on Urban Sustainability. London: James & James

Healey, P. 1997. Collaborative Planning: Shaping Places in Fragmented Societies. London: Macmillan

Hoyle, B.S. et al (ed.) 1988. Revitalising the Waterfront. London: Belhaven

IBA. 1999. International Bauausstellung Emscher Park. Wuppertal

ICLEI. 2002. Local Strategies to Accelerate Sustainability. Ottawa

ICLEI. 1997. From Charter to Action. Report of the Second European Conference on Sustainable Cities and Towns. Lisbon, October 1996

ICLEI. 1995. Towards Sustainable Cities and Towns. Report of the First European Conference on Sustainable Cities and Towns. Aalborg, May 1994

IEA. 2008a. World Energy Outlook 2008. Paris

IEA. 2008b. Deploying Renewables. Principles for Effective Policies. Paris

IEA. 2004a. Renewable Energy Market and Policy Trends in IEA. Paris

IEA. 2004b. Biofuels for Transport. An International Perspective. Paris

IEA. 2003a. Energy to 2050. Scenarios for a Sustainable Future. Paris

IEA. 2003b. Renewables for Power Generation. Paris

International Institute of the Urban Environment. 1995. Environmental Awareness Workshops. The Hague

INU – Politecnico di Milano. 1997. Il Tempo e la città tra natura e storia. Atlante di progetti sui tempi della città. Milan

IPCC. 2007. Fourth Assessment Report, Climate Change 2001. Summary for Policy Makers. Bonn

International Transport Forum. 2009. Transport For A Global Economy. Challenges Opportunities in the Downturn. Leipzig

Jacobs, J. 1985. La Ville et la richesse des nations. Réflexions sur la vie economique. Quebec: Boréal

Jacobs, J. 1969. The Economy of Cities. London: Penguin

Jimenez, J. 2002. Teoria de arte. Madrid: Tecnos

Koolhas, R. 1995. What Ever Happened to Urbanism ? S.M.L.XL. Rotterdam: 010

Landry, C. and Bianchini, F. 1995. The Creative City. London: Demos/Comedia

Le Corbusier. 1971. La Charte d'Athènes. Paris: Éditions du Seuil

Leeds Metropolitan University. Centre pour le développement urbain et la gestion de l'environnement. 2007. Compétences pour le futur 2006. Leeds

Le Galès, P. 2001. Le retour des villes européennes. Paris: Grasset

Malmö. 2008. Making Sustainability Reality. Malmö Office of City Planning

Martinotti, G. 1993. Metropoli, la nuova morfologia sociale. Bologna: Il Mulino

Mc Donald F. 1989. Saving the City. How to halt the destruction of Dublin. Dublin: Tomar

Mega, V. 2008. Modèles pour les villes d'avenir. Un kaléidoscope de modèles pour des villes durables. Paris: L'Harmattan

Mega, V. 2007. Towards the civilisation of sustainable energy. The European perspective in Regional Symbiosis, Vol 12, 13/14, Kanpur, India

Mega, V. 2005. Sustainable Development, Energy and the City. New York: Springer

Mega, V. 2004. *Urban Dimensions of Sustainable Development* (featured author) in 1.18. Human Resource System Challenge VII: Human Settlement Development in Encyclopedia of Life Support Systems,.UNESCO. Oxford: EOLSS

Mega, V. 2000. *Innovations for the civilisation of sustainability. Harmonising policy objectives in the European urban archipelago* In Proceedings 2nd Biennial of European cities and Town planners. Berlin

(The) Megacities Foundation. 2008. Towards the Megacities Solution. Report of the proceedings of the Jubilee Congress. Delft. November 2008

Mercer. 2009a. Quality of Living Survey. New York

Mercer. 2009b. Cost of Living Survey. New York

Mercer. 2008. Quality of Living Survey. New York

Mercer. 2007. Quality of Living Survey. New York
METROPOLIS. 1999. A Network of Cities for World Citizens. 6th World Congress documents. Barcelona
METROPOLIS. 1996. Metropolis for the People. 5th World Congress Documents. Tokyo
M.I.T. 2008. Architecture for the Carbon-free City. Policy seminar June 2008
M.I.T. 1997. Promoting Innovation. Summer Seminar (organised by S. Weiner). Working papers. Cambridge
Monod, J. 2009. Les vagues du temps. Mémoires. Paris: Fayard
Municipality of Amsterdam. 1994. Car-Free Cities ? Working Papers
Municipality of Athens. 1994. From the Organic City to the City of Citizens. Athens
Munier, N. 2007. Handbook on Urban Sustainability. New York: Springer
Neal, P. 2003. Urban Villages and the Making of Communities. London: Spon
Niemeyer, O. 1997. Les courbes du temps. Paris: Gallimard
Nijkamp, P. and Perrels, A. 1994. Sustainable Cities in Europe. London: Earthscan
Nuvolati, G. 1998. La qualità della vita delle città. Milan: Angeli
Obama, B. 2009. Next Steps in the Development of Urban and Metropolitan Agenda. Washington: The White House
OECD. 2009a. Local Responses to a Global Crisis. International Strategies for Recovery
OECD. 2009b. Doing Better for Children. Paris
OECD. 2008a. Environmental Outlook and Strategy. Paris
OECD. 2008b. OECD Forum on Climate Change, Growth and Stability. Paris
OECD. 2007a. Competitive Cities. A New Entrepreneurial Paradigm in Spatial Development. Paris
OECD. 2007b. OECD Forum on Innovation, Growth and Equity. Paris
OECD. 2006. Competitive Cities in a Global Economy. Territorial Reviews. Paris
OECD. 2005. OECD Forum on Fuelling the Future (Security, Stability, Development)
OECD. 2004. Implementing Sustainable Development. Key Results 2001-2004. Paris
OECD. 2003. Poverty and Climate Change. Reducing the Vulnerability of the Poor through Adaptation. Paris
OECD. 2002. Governance for Sustainable Development. Paris
OECD. 2001a. Policies to Enhance Sustainable Development. Paris
OECD. 2001b. OECD Forum on Sustainable Development and the New Economy. Highlights. Paris
OECD. 2000a. Towards Sustainable Development: Indicators to Measure Progress. Paris
OECD. 2000b. Cities for Citizens: Improving Governance in Metropolitan Areas. Paris
OECD. 2000c. The Kyoto Protocol and Beyond. Paris
OECD. 1996. The Ecological City. Paris
OECD - ECMT. 1994. Urban Travel and Sustainable Development. Paris
OECD - Toronto. 1997. Better Governance for More Competitive and Liveable Cities. Conference Documents, Toronto
Olsen, D. 1987. La città come opera d'arte. Milan: Sena e Riva
O'Sullivan, A. 1996. Urban Economics. Chicago: Irwin
Owen, D. 2009. Green Metropolis. New York: Riverhead Books
Parkinson, M. 1998. Combating Social Exclusion: Lessons from area-based Programmes in Europe. London: Policy
Petrella, R. 2001. The Water Manifesto: Arguments for a World Water Contract. Zed Books
Pistor et al. 1994. A City in Progress. Physical planning in Amsterdam. Amsterdam
Putnam, R. et al. 2002. Democracies in Flux: The Evolution of Social Capital in Contemporary Society. Oxford University Press
Roaf, S. 2007. Ecohouse. Third edition. London: Elsevier
Rowe, P. 2006. Building Barcelona. A second Renaissance. Barcelona: Actar
Région de Bruxelles-Capitale. 1995. Manuel des espaces publics Bruxellois. Brussels: Iris
Renner, M et al. 2009. Toward a Transatlantic Green New deal: tackling the Climate and Economic challenges. Washington: Worldwatch Institute and Heinrich Böll Foundation.

Rogers, R and Gumudjian, P. 1994. Cities for a Small Planet. London. Westview

Rogers, R. and Urban Task Force. 2005. Towards an Urban Renaissance. London: The Mayor's Office

Rueda, S. 1995. Ecologia Urbana. Barcelona: Beta Editorial

Sachs, J. 2008. Common Wealth. Economics for a Crowded Planet. New York: Penguin Books

Sansot, P. 1973. La poétique de la ville. Paris: Klincksieck

Sapolsky, H.M. 1990. The politics of risk. Dedalus. Cambridge

Sassen, S. 1994. Cities in a World Economy. London: Pine Forge

Sassen, S. 1991. The Global City: New York, London, Tokyo. Princeton Univeryity Press

Schmidt-Eichstaedt, G. 1993. Schwabach, Modell Stadt - Okologie. Berlin: TU Institut für Stadt und Regionalplannung

Schumpeter, J. 1976. Capitalism, Socialism and Democracy. New York: Harper and Row

Sennet, R. 1990. The Conscience of the Eye. The Design of Social Life of Cities. New York: Knopf

Sitte, C. 1889. Der Städtebau nach seinen künstlerischen GrundsätzenCity. Traduit en L'art de bâtir les villes. L'urbanisme selon ses fondements artistiques. Paris: L'équerre

Stern, N. 2009. A Blueprint for a Safer Planet. London: Bodley Head

Stern, N. 2006. The economics of Climate Change. Cambridge University Press

Stiglitz, J. 2009. The Great GDP Swindle. In Guardian. London

Sustainable Seattle. 2008. Indicators of Sustainable Community. Seattle

SustainAbility/UNEP. 1998. Engaging Stakeholders. The non-Reporting report. London

Tetraplan A/S (coord.). 2009. Transvisions. Report on Transport scenarios with a 20 and 40 year horizon. Copenhagen

Tombazis, A. 2007. Letters to a Young Architect. Athens

Touraine, A. 1997. Pourrons-nous Vivre Ensemble ? (Egaux et Différents). Paris: Fayard

UITP. 2007. Making Tomorrow Today. Brussels

UITP. 2005. Bringing Quality to Life. Brussels

UITP. 2004. Ticket to the Future: 3 Stops to Sustainable Mobility. Brussels

UNITED NATIONS. 2009. The UN Population Prospects: the 2008 Revision Population Database, New York

UNITED NATIONS. 2007. UN World Urbanization Prospects: The 2007 Revision Population Database, New York

UNITED NATIONS. 2006. The UN Population Prospects: The 2006 Revision Population Database, New York

UNITED NATIONS. 2005. The Millennium Development Goals Report. New York

UNITED NATIONS. 2004. World Urbanisation. The 2003 Revision. New York

UNITED NATIONS. 2002. Financing for Development. Adoption of the Monterrey Consensus. New York

UNITED NATIONS. 2001. Cities in a Globalizing World: Report on Human Settlements 2001, New York

UNITED NATIONS. 1997. Report on the Summit on Human Settlements. New York

UNITED NATIONS. 1996. Indicators of Sustainable Development. Framework and Methodology. New York

UNAIDS. 2008. 2008 Report on the Global AIDS Epidemic. Geneva

UNDP. 2009. Human Development Report 2009. New York

UNDP/UNCHS (Habitat) / World Bank (Urban Management Programme). 1993. Environmental Innovation and Management in Curitiba, Brazil

UNECE. 2002. Sustainable and Liveable Cities. Leeds Conference Documents. Geneva

UNEP. 2008. Green Jobs: Towards Sustainable Work in a Low-Carbon World

UNEP. 2000. Global Environmental Outlook 2000. New York: Earthscan

UNEP/HABITAT. 2005. Urban Air Quality Management Toolbook. Nairobi

UNEP/ICLEI/Cities Alliance. 2007. Liveable Cities. The Benefits of Urban Environmental Planning. Washington

UNEP/ILO. 2008. Green Jobs. Towards Decent Work in a Sustainable, Low-Carbon World. Nairobi, Geneva

UNESCO. 2004. Encyclopaedia of Life Support Systems. Oxford: EOLSS

UNESCO (COMEST Sub-Commission). 2001. The Ethics of Energy. A Framework for Action. Paris

UNESCO. 1995. Les Libertés de la Ville. Paris: Passages

UNESCO. 1988. MAB, Towards the Sustainable City. Paris

UN/HABITAT. 2007. Enhancing Urban Safety and Security. Global report on Human Settlements 2007. Nairobi

UN/HABITAT. 2006. State of the World's Cities 2006/7. Nairobi

UN/HABITAT. 1996. An Urbanizing World, Global Report on Human Settlements. Oxford: Oxford University Press

UNFPA. 2004. State of World Population. New York

UN / Tokyo Metropolitan Government. 1998. Eco - partnerships Tokyo. Cultivating an Eco-Society. Tokyo

Veolia Environment. 2008. Observatoire Veolia des modes de vie urbains 2008. L'état de la vie dans les villes. Etude réalisée par Ipsos. Paris

World Bank. 2009. World Development Report. Washington D.C.

World Bank. 1995. The Human Face of the Urban Environment. Washington

WBCSD. 2009. Corporate Ecosystem Valuation - Building the Business Case. Geneva

WBCSD. 1998. Signals of Change. Geneva

WHO. 2008. City Leadership for Health. Summary Evaluation of Phase IV of the WHO European Healthy Cities Network. Copenhagen

WHO. 2003. International Healthy Cities Conference. Conference Documents. Belfast

WHO – OECD. 1996. Our Cities, Our Future. Proceedings of the First International Congress on Healthy and Ecological Cities. Copenhagen

Wood, A. 2008. Best Tall Buildings of the World. Brussels

Wood, J. 2008. Local Energy: Distributed Generation of Heat and Power. London

World Economic Forum. 2008. Innovation 100. Stanford University

World Economic Forum. 2007. Global Gender Gap Report. Geneva

World Resources Institute (WRI). 2006. Hot Climate, Cool Commerce: A Service Sector Guide to Greenhouse Gas Management. Washington

Worldwatch Institute. 2009a. Vital signs 2009. Washington

Worldwatch Institute. 2009b. State of the World 2009: Into a warming world. Washington

Worldwatch Institute. 2008a. Green Jobs: Working for People and the Environment. Washington

Worldwatch Institute. 2008b. State of the World 2008: Innovations for a sustainable economy. Washington

Worldwatch Institute. 2007. State of the World 2008: Our Urban Future. Washington

Zayczyk, F. 2000. Tempi di vita e orari della città: la ricerca sociale e il governo urbano. Milan: Angeli

Index

LaVergne, TN USA
18 March 2011
220734LV00006B/22/P